V. P. Dimri

Fractal Behaviour of the Earth System

V. P. Dimri (Editor)

Fractal Behaviour of the Earth System

Foreword by Prof. D. L. Turcotte

With 172 Figures, 19 in colour

 Springer

EDITOR:

DR. V. P. DIMRI
NATIONAL GEOPHYSICAL RESEARCH
INSTITUTE
UPPAL ROAD
HYDERABAD, 500 007

INDIA

E-MAIL: VPDIMRI@NGRI.RES.IN

COVER PAGE: TSUNAMI WAVE GENERATED USING FRACTAL ART BY KEN KELLER, REF: HTTP://FRACTALARTGALLERY.COM

ISBN 13   978-3-642-06585-9
e-ISBN 13   978-3-540-26536-8

Springer is a part of Springer Science+Business Media
springeronline.com
© Springer-Verlag Berlin Heidelberg 2010
Printed in The Netherlands

Cover design: E. Kirchner, Heidelberg

Printed on acid-free paper   32/2132/AO 5 4 3 2 1 0

Foreword

It is with pleasure that I write the foreword to this excellent book. A wide range of observations in geology and solid-earth geophysics can be explained in terms of fractal distributions. In this volume a collection of papers considers the fractal behavior of the Earth's continental crust. The book begins with an excellent introductory chapter by the editor Dr. V.P. Dimri. Surface gravity anomalies are known to exhibit power-law spectral behavior under a wide range of conditions and scales. This is self-affine fractal behavior. Explanations of this behavior remain controversial.

In chapter 2 V.P. Dimri and R.P. Srivastava model this behavior using Voronoi tessellations. Another approach to understanding the structure of the continental crust is to use electromagnetic induction experiments. Again the results often exhibit power law spectral behavior. In chapter 3 K. Bahr uses a fractal based random resister network model to explain the observations.

Other examples of power-law spectral observations come from a wide range of well logs using various logging tools. In chapter 4 M. Fedi, D. Fiore, and M. La Manna utilize multifractal models to explain the behavior of well logs from the main KTB borehole in Germany. In chapter 5 V.V. Surkov and H. Tanaka model the electrokinetic currents that may be associated with seismic electric signals using a fractal porous media.

In chapter 6 M. Pervukhina, Y. Kuwahara, and H. Ito use fractal networks to correlate the elastic and electrical properties of porous media.

In chapter 7 V.P. Dimri and N. Vedanti consider fractal distributions of thermal conductivity in determining the thermal structure of the lithosphere. In chapter 8 L. Telesca and Vincenzo Lapenna apply fractal concepts to streaming potentials in porous media. And in chapter 9 H.N. Srivastava considers the embedding dimensions of seismicity patterns.

The editor, Dr. V.P. Dimri, is to be congratulated on putting together an excellent collection of related papers. The basic theme is the use of fractal techniques to better understand the geophysics of the continental crust. The papers included are important for both fundamental and applied reasons.

October, 2004   *Donald L. Turcotte*

Department of Geology, University of California, USA

Dedicated to the Victims
of
26 December 2004, Indian Ocean Tsunami

Preface

Though the theory of fractals is well established, but still it is in nascent stage of its widespread applicability in geosciences. There is clear-cut hiatus between theoretical development and its application to real geophysical problems in the available literature. The present book is an attempt to bridge this existing gap and it is primarily aimed at postgraduates and researchers having interest in geosciences.

In this book the state-of-the-art fractal theory is presented and the reader obtains an impression of the variety of fields for which scaling and fractal theory is a useful tool and of the different geophysical methods where it can be applied. In addition to the specific information about multi-geophysical applications of fractal theory in imaging of the heterogeneous Earth, ideas about how the theory can be applied to other related fields has been put forward. A strong point for writing this book is to touch almost all the topics of geophysics used for oil/mineral exploration viz. potential field methods, electrical methods, magnetotellurics and fractals in geothermics (a new concept proposed first time), which are very important topics for any one practicing geoscience. The chapters are written by internationally known Earth science experts from Germany, Italy, Russia, Japan, and India with numerous references at the end of each chapter.

This is the first book of its kind, contributed by some of the pioneers in the related fields, which covers many applications of fractal theory in multi-geophysical methods, at an accessible level. The complicated concepts are introduced at the lowest possible level of mathematics and are made understandable. Readers will find it as an interesting book, to read from cover to cover.

I take this opportunity to express my sincere thanks and regards to Professor Donald L. Turcotte for reading the first draft of the chapters and writing the foreword for this book.

Finally, I would express my deep gratitude towards Professors Karsten Bahr, Maurizio Fedi, Vadim V. Surkov, Drs Marina Pervukhina, Luciano Telesca, H.N. Srivastava, and all other contributing authors, who have cooperated with me at every stage of editing this volume. We thank all the publishers and authors for granting permission to publish their illustrations and tables in the volume.

Last but not least my special thanks are due to wife Kusum, who supported my endeavor with great patience.

February 11, 2005   **V.P.Dimri**

Contents

List of contributors

Bahr Karsten
Institut für Geophysik,
Herzberger Landstr. 180, 37075 Göttingen, Germany
E-mail: kbahr@geo.physik.uni-goettingen.de

Dimri V.P.
National Geophysical Research Institute,
Uppal Road, Hyderabad, 500 007, India
E-mail: vpdimri@ngri.res.in

Fedi Maurizio
Università di Napoli Federico II
Dipartimento di Scienze della Terra
Largo san Marcellino 10 80138, Napoli - Italia
E-mail: fedi@unina.it

Fiore Donato
Università di Napoli Federico II
Dipartimento di Scienze della Terra
Largo san Marcellino 10 80138, Napoli - Italia
E-mail: donatofiore@aliceposta.it

Ito Hisao
Geological Survey of Japan AIST,
Higashi 1-1-3, Tsukuba, Ibaraki 305-8567, Japan
E-mail: hisao.itou@aist.go.jp

Kuwahara Yasuto
Geological Survey of Japan AIST,
Higashi 1-1-3, Tsukuba, Ibaraki 305-8567, Japan
E-mail: y-kuwahara@aist.go.jp

La Manna Mauro
Università di Napoli Federico II
Dipartimento di Scienze della Terra
Largo san Marcellino 10 80138, Napoli - Italia
E-mail: lamanna@unina.it

Lapenna Vincenzo
Institute of Methodologies for Environmental Analysis,
National Research Council, C.da S.Loja, 85050 Tito (PZ), Italy
E-mail: lapenna@ imaa.cnr.it

Pervukhina Marina
Geological Survey of Japan AIST,
Higashi 1-1-3, Tsukuba, Ibaraki 305-8567, Japan
E-mail: marina-pervuhkina@aist.go.jp

Srivastava H. N.
128, Pocket A, Sarita Vihar,
New Delhi, 110 044
E-mail: hn_srivastava@hotmail.com

Srivastava Ravi Prakash
National Geophysical Research Institute,
Uppal Road, Hyderabad, 500 007, India
E-mail: ravi_prakash@ngri.res.in

Surkov Vadim V.
Moscow State Engineering Physics Institute,
115409 Moscow, Kashirskoe road, Russia
E-mail: surkov@redline.ru

Tanaka H.
RIKEN, International Frontier Research Group on Earthquakes,
c/o Tokai University, Shimizu, Japan
E-mail: tanaka4348@yahoo.co.jp

Telesca Luciano
Institute of Methodologies for Environmental Analysis,
National Research Council, C.da S.Loja, 85050 Tito (PZ), Italy
E-mail: ltelesca@imaa.cnr.it

Vedanti Nimisha
National Geophysical Research Institute,
Uppal Road, Hyderabad, 500 007, India
E-mail: nimisha@ngri.res.in

Chapter 1. Fractals in Geophysics and Seismology: An Introduction

V.P.Dimri

National Geophysical Research Institute, Hyderabad, India

1.1 Summary

Many aspects of nature are very much complex to understand and this has started a new science of geometrical complexity, known as 'Fractal Geometry'.Various studies carried out across the globe reveal that many of the Earth's processes satisfy fractal statistics, where examples range from the frequency-size statistics of earthquakes to the time series of the Earth's magnetic field. The scaling property of fractal signal is very much appealing for descriptions of many geological features. Based on well-log measurements, Earth's physical properties have been found to exhibit fractal behaviour. Many authors have incorporated this finding in various geophysical techniques to improve their interpretive utility. The aim of present chapter is to briefly discuss the fractal behaviour of the Earth system and the underlying mechanism by citing some examples from potential field and seismology.

1.2 Introduction

B.B. Mandelbrot, in his book "Fractal Geometry of Nature" writes that the fractal geometry developed by him, describes many of the irregular and fragmented patterns around us, and leads to full-fledged theories, by identifying a family of shapes, he calls "Fractals". Using this geometry, complex patterns of nature like rocky coast lines, shape of the clouds, jagged surfaces of mountains etc. can be mapped.

The classical geometry deals with objects of integer dimensions. Zero dimensional points, one dimensional lines, two dimensional planes like squares, and three dimensional solids such as cubes make up the world as we have previously understood it, but the newly coined fractal geometry describes non-integer dimensions. Many natural phenomena are better described with a dimension partway between two whole numbers. So while a straight line has a dimension of one, a fractal curve will have a dimension between one and two depending on how much space it takes up as it twists and curves (Peterson 1984). The more that flat fractal fills a plane, the closer it approaches two dimensions. So a fractal landscape made up of a large hill covered with tiny bumps would be close to the second dimension, while a rough surface composed of many medium-sized hills would be close to the third dimension (Peterson 1984). A higher fractal dimension means a greater degree of roughness and complexity, for example the smooth eastern coast of Florida has a fractal dimension very close to unity, while the very rugged Norwegian coast having fjords has fractal dimension $D = 1.52$. Fractal geometry is a compact way of encoding the enormous complexity of many natural objects. By iterating a relatively simple construction rule an original simple object can be transformed into an enormously complex one by adding ever increasing detail to it, at the same time preserving affinity between the whole and the parts or scale invariance, which is very significant property of fractals. The essence of fractal theory lies in the scaling of properties.

Mandelbrot and Van Ness (1968) extended the concept of fractals in terms of statistical self-similarity or scale invariance in time series analysis which was done within the context of self-affine time series. The basic definition of a self affine time series is that the power spectrum has power-law dependence on frequency.

Fractal concept is very much useful for interpretation of time series data in various branches of Earth science like horizontal variability of temperature, humidity, rainfall, cloud water in atmosphere etc. All these phenomena obey power law behavior over well-defined wavenumber ranges. These results are very much important for understanding the variability of the atmosphere and for improved characterization of these fields into large scale models of the climate system. A number of properties of the solid Earth have been discussed in fractal terms (Turcotte 1992). Fractal theory in geophysical observations has numerous applications in correlating and predicting situation from known to unknown and hence has attracted the attention of geoscientists.

1.3 Fractal signal analysis

Geophysical data are in the form of time series. A geophysical time series can be characterized by combination of stochastic component, trend component and periodic component (Malamud and Turcotte 1999). To quantify the stochastic component of the time series it is necessary to specify the statistical distribution of values and persistence. There exists variety of techniques to quantify the strength of persistence, but the most commonly used is the spectral analysis, where the Fourier spectrum of time series is plotted against frequency (or wavenumber in case of space series) and the value of slope, known as scaling exponent (say β) gives an estimation of persistence. Depending upon the value of scaling exponent β persistence can be characterized as weak or strong.

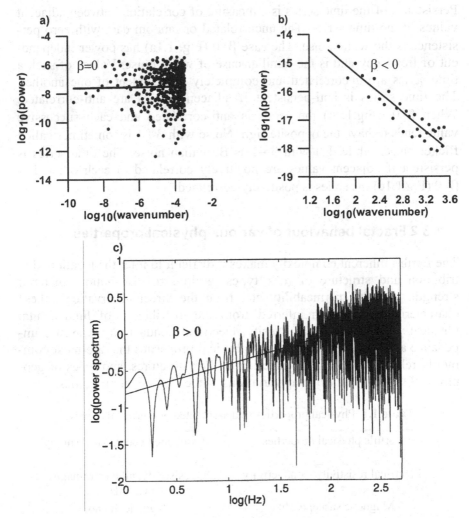

Fig. 1.1 Power spectrum of (a) random data, (b) persistent data, and (c) anti-persistent data

Scaling exponent controls the balance of high and low frequencies and determines the degree of smoothness or correlation of the series. There exists a relation between the scaling exponent, Euclidean dimension (E) and fractal dimension (D) of data, given as D=E-1-β.

1.3.1 Analyzing geophysical time series

Persistence of the time series is a measure of correlation between adjacent values of the time series. The uncorrelated or random data with zero persistence is the white noise. The case β=0 (Fig. 1.1a) has power independent of frequency and is the familiar case of white noise. Values of such a time series are uncorrelated and completely independent of one another. The time series is anti-persistent if adjacent values are anti-correlated. When β<0, (Fig.1.1b) the series is anti-correlated and each successive value tends to have the opposite sign. Noise with β=-1 is sometimes called flicker noise, while that with β=-2 is Brownian noise. The time series is persistent if adjacent values are positively correlated to each other. For β>0 (Fig.1.1c) the series is positively correlated.

1.3.2 Fractal behaviour of various physical properties

The Earth's inherent complexity makes it difficult to infer the location, distribution and structure of rock types, grain size distribution, material strength, porosity, permeability etc. from the direct observations; these characteristics are often inferred from the distribution of fundamental physical properties such as density, electrical conductivity, acoustic impedance and others. Table 1.1 lists physical properties that are most commonly related to geological materials and/or structures, and types of geophysical surveys that can map variations of these physical properties.

Table 1.1 Physical properties and associated geophysical surveys

Earth's physical properties	Associated geophysical survey
Electrical resistivity /conductivity	DC resistivity, all electromagnetic methods
Magnetic susceptibility	Magnetic methods
Density	Gravity, and Seismic methods
Acoustic wave velocity	Seismic reflection or refraction

Other physical properties that can be usefully mapped include charge-ability, natural radioactivity, dielectric permittivity, and porosity. These properties, measured indirectly through geophysical surveys, record the Earth's response. It is observed from the German Continental Deep Drilling Programme (KTB) examples illustrated in Fig. 1.2 that the source distribution of above-mentioned physical properties follow power-law, hence they are fractal in nature.

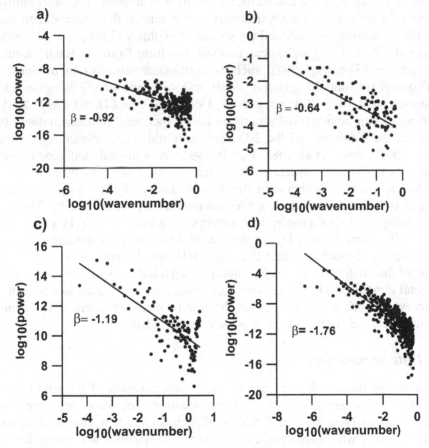

Fig.1.2 Power spectrum for (a) density, (b) susceptibility, (c) electrical resistivity and (d) thermal conductivity data from KTB borehole Germany

For geophysical study the physical response of the Earth can be approximated by convolution model (Dimri 1992). Extraction of useful information from the observations made over the surface needs advanced processing and interpretation techniques. Fractal theory finds vivid applications in all the aspects of geophysical exploration, which has been discussed in subsequent section by citing some examples mainly from acquisition, processing and interpretation of potential field data.

Data acquisition

Geophysical surveys are generally carried out along the existing roads. Such a convenient survey can miss an anomaly of interest. An optimal design of survey network can delineate anomaly of interest. The detectibility limit of a large-scale geophysical survey depends on the fractal dimension of the measuring network and the source of anomaly (Lovejoy et al. 1986; Korvin 1992). The geophysical anomaly resulting from the fractal nature of sources (Turcotte 1992), such as nonrandom distribution of density (Thorarinsson and Magnusson 1990) and susceptibility (Pilkington and Todoeschuck 1993, Pilkington et al. 1994, Maus and Dimri 1994; 1995; 1996), cannot be measured accurately unless its fractal dimension does not exceed the difference of the 2-D Euclidean and fractal dimension of the network (Lovejoy et al. 1986, Korvin 1992). A well-designed geophysical survey can delineate structures of interest that otherwise would be missed. A theoretical relation between the dimensionality of geophysical measuring network and anomaly has been established by Dimri (1998). The detectibility limits of geophysical surveys is given by Korvin (1992) as $D_s = E - D_n$ where D_n and D_s are the fractal dimensions of the network and the source, respectively, and E is the Euclidean dimension. Sources with fractal dimensions less than D_s cannot be detected by a survey network of fractal dimension D_n. Hence, the fractal concept can be used to design the survey network in order to detect the small sources of interest. The same concept can be applied to design seismic arrays also.

Data processing

The fractal theory plays a crucial role in interpolation of the data at the time of processing, as the spatial locations of data sets are inhomogeneously distributed. Gridding of such data suffers from the interpolation errors, which are manifested in terms of spurious anomalies due to aliased-interpolated data. Short wavelength anomalies of potential field data produce aliasing, which can be minimized using this approach. The fractal dimension of measuring network characterizes the data distribution and represents the density of data distribution in simplest way, unlike other techniques. Using fractal dimension, optimum gridding interval can be obtained, which is used for optimum interpolation interval obeying Shanon's sampling theorem. This concept is being widely used while processing of potential field data where an error in gridding and interpolation can lead to spurious results.

The potential field data are often subject to a number of processing techniques to improve their interpretive utility, which can be efficiently carried out in the frequency domain using fast Fourier transform (FFT). In general the acquired geophysical data sets do not satisfy basic requirements of FFT, hence grid extrapolation is required by filling in any undefined values and enlarging the original grid size so that the effects of the periodicity is minimized. In such cases a fractal based conditional simulation method (Tubman and Crane 1995) can be used for grid extrapolation.

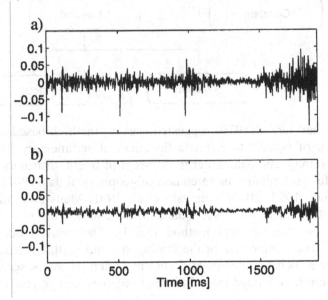

Fig. 1.3 The error between the reflectivity and the output from (a) standard deconvolution and (b) scaling deconvolution method. The relative error energy is 13% for standard deconvolution, and 5% for scaling deconvolution (after Toverud et al 2001)

In seismic data processing deconvolution plays a crucial role and to incorporate the fractal behaviour of reflectivity sequence, new deconvolution operators have been designed (Todoeschuck and Jensen 1988; 1989, Saggaf and Toksoz 1999). Toverud et al (2001) have compared the performance of standard deconvolution with scaling deconvolution and arrived to the conclusion that the percentage error involved with recovery of reflectivity series using fractal based scaling deconvolution is less than that compared to the standard deconvolution which is optimal only for a white noise reflectivity (Fig. 1.3).

Data interpretation

Interpretation of geophysical data is usually carried out in space and frequency domain. Normally one of the following types of source distribution is assumed in geophysical interpretation.

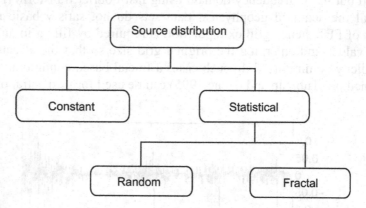

Spector and Grant (1970) suggested spectral method, based on random distribution of source, to estimate thickness of sedimentary basins from gravity and magnetic data. Later the concept of fractal distribution was introduced for preliminary interpretation of geophysical data (Pilkington and Todoeschuck 1990; 2004, Gregotski et al. 1991, Maus and Dimri 1994; 1995; 1996, Fedi et al. 1997, Quarta et al. 2000). This new method is known as scaling spectral method (SSM). The conventional spectral method is a particular case of the scaling spectral method. Second order statistics that includes the power spectrum of density and susceptibility of the core samples obtained from different boreholes, can be given as

$$P(f) = f^{-\beta} \qquad\qquad (1.1)$$

where P(f) is power spectrum, f is wavenumber and β is the scaling exponent. For mathematical convenience, earlier authors have assumed random behavior of sources ($\beta=0$). The SSM has been applied to many potential field data sets (Fig. 1.4) for estimation of depth to the top of the sources. Bansal and Dimri (1999; 2001) demonstrated the application of this method to interpret gravity and magnetic data acquired along the Nagur-Jhalawar and Jaipur-Raipur geo-transacts in India.

Fractal models can be used in the inversion of potential field data. Computation of forward gravity response over a fractal structure has been described in Chapter 2 of this book. In case of magnetics, when the data is inverted by using models of constant susceptibility distribution blocks, the fractal behaviour can be introduced through the parameter covariance ma-

trix, in the least squares solution of the problem (Pilkington and Todoeschuck 1991). The parameter covariance matrix controls the smoothness of the final solution. The matrices can be designed to maximize the
smoothness, under the assumption, that the results contain a minimum
amount of features that are found more likely to represent those which are
actually present in the data (Constable et al. 1987). Pilkington and Todoeschuck (1991) showed that the inclusion of fractal behaviour of variables in the inversion is equivalent to imposing a smoothness constraint.
Such inversions can be called naturally smooth on the grounds that they
have their smoothness determined in an objective manner from data.

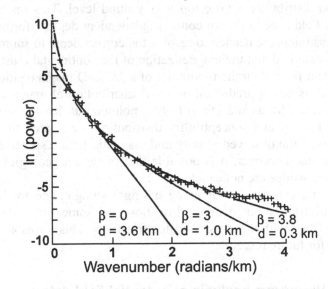

Fig. 1.4 Power spectrum of aeromagnetic data over a sedimentary basin (data
from Pilkington et al. 1994). The correct depth to source is approximately 1700 m.
The solid lines indicate the best fit of the model power spectrum in a least square
sense for some selected values of β. For β =3.8 the best overall fit is obtained,
β=3.0 is the mean value of of the scaling exponent for aeromagnetic data derived
by Gregotski et al. (1991) and β=0 corresponds to the Spector and Grant method.
The smaller the values we assume for β, greater the estimated depth to source. (after Maus and Dimri 1996)

The effect of including fractal behaviour in magnetic susceptibility for
the inversion of magnetic data has been illustrated by Pilkington (1997)
with a 2-D example using a conjugate gradient inversion algorithm. Results obtained using β=0, showed spurious variations in susceptibilities,
which were likely to be caused by the inversion method itself. In contrast

of above, the scaling solution behaved well and had a level of smoothness, consistent with the prior knowledge of the spatial variability of the variable for which it was solved.

Another practical example is the use of aeromagnetic data to estimate the depth to basement in sedimentary basins. As sedimentary rocks generally have low susceptibilities the field can be taken as originating from crystalline rocks of the basement. Downward continuation of a white power spectrum greatly overestimates the source depth (Pilkington et al. 1994). Maus and Dimri (1996) have examined potential field spectra and found that in each case, the spectra can be fit well simply by a fractal magnetization distribution whose top is at ground level. This reveals that the potential field spectra do not contain independent depth information and a priori constraints are needed to resolve the correct depth to source.

If it is assumed that the magnetization of the continental crust is induced by the main field, then the prediction of a $1/f^4$ 3-D magnetization distribution (field) is also a prediction of a $1/f^4$ distribution of magnetic susceptibility (source). Maus and Dimri (1994) pointed out that the scaling exponents of density and susceptibility distribution are related to the scaling exponents of the observed gravity and magnetic field respectively by simple equations. However, it is possible that in certain geological situations these relationships are not exact.

Here, one can assert that using a single straight line to describe the power spectrum is an oversimplification and some times the spectrum could be fitted with more complicated functions. This forms an interesting problem for future research.

1.3.3 Variogram analysis of potential field data

Maus (1999) claimed that variograms are the appropriate space domain statistical tools to analyze magnetic and possibly gravity data. He transformed a self similar spectral model analytically to the space domain in order to avoid the distorting effects of transforming the measured data to wavenumber domain. After describing the gravity and magnetic scaling spectral models Maus (1999), has derived the corresponding variogram model for the complex case of aeromagnetic profiles in a non-vertical inducing field. The variogram model for gravity data is subsequently derived as a special case.

It is noteworthy to mark that model variograms do depend on scaling exponent besides other related parameters viz. orientation of the profile (in magnetic case), depth, and source intensity. Hence, the precise estimation of scaling exponent is essential for determination of depths using the variogram analysis.

1.4 Analysis of seismological data

1.4.1 Size-Frequency distribution of earthquakes

The relationship between the size of an earthquake and its frequency of occurrence obeys fractal statistics. It is well known that very large earthquakes are rare events and very small earthquakes are very frequent (seismically active regions can register hundreds of small earthquakes per day). What is remarkable is that there is a power law relationship between the number of large and number of small earthquakes in a given region per unit time. For instance every year, globally there is on an average just one earthquake of magnitude eight, ten earthquakes of magnitude seven, one hundred earthquakes of magnitude six, one thousand earthquakes of magnitude five and so on. The power law in this case is

$$N(m) = a \ A^{-b} \qquad (1.2)$$

where N(m) is the total number of earthquakes per unit time in a given region with magnitude m or greater, and A is the amplitude of ground motion, and $m \sim \log(A)$. Taking logarithm, the power law becomes

$$\log N \ (>m) = \log (a) - bm \qquad (1.3)$$

This empirical formula is known as the Gutenberg-Richter law of earthquake. Globally the value of b (usually called the "b-value") is observed to be around 1.0, and the constant 'a' is about 1×10^8 /year (it represents the number of magnitude 1 earthquakes in a year). The relation between D (fractal dimension) and b is given as

$$D = 3b/c \qquad (1.4)$$

where c is a constant depending on the relative duration of the seismic source and time constant of the recording system. For crystalline rocks the value of c is taken to be 3.0, for subduction zones (100-700 km depth) this value is suggested to be 2.4 and for most of the earthquake studies it is believed to be 1.5 (Kanamori and Anderson 1975).

1.4.2 Omori's law: analysis of aftershock data

Omori's law (1894) in seismology is recognized as the first fractal law in seismology (Moharir 2000). According to it, the rate of aftershock events decays as the inverse power law in elapsed time after the main shock. The Omori's law describes the decay of aftershock activities with time. It is characterized by a power law given as

$$N(t) \sim t^{-p} \tag{1.5}$$

where, $N(t)$ is the number of aftershock events in unit time interval after the main shock and 'p' is the rate of decay of aftershocks. In Omori law, the p-value ranges from 0.5 to 2.5 but normally comes closer to 1. Subsequently the Omori's law has been modified. The modified Omori law describes the rate of decay of an aftershock sequence by equating the number of aftershocks at some time t after a mainshock with the quantity t plus K_1 (a constant) to the negative power of 'p', all multiplied by K_2 (another constant):

$$N(t) = K_2 * (t + K_1)^{-p} \tag{1.6}$$

1.4.3 Fractal dimension and seismicity distribution

The fractal theory has led to the development of a wide variety of physical models of seismogenesis including nonlinear dynamics and it can be efficiently used to characterize the seismicity pattern of a region. The fractal nature of the spatial distribution of earthquakes was first demonstrated by Kagan and Knopff (1980), Hirata and Imoto (1991), and Hirabayashi et al. (1992). The hypocenter distribution data suggests that the changes in fractal dimension could be a good precursor parameter for earthquakes as it is a measure of the degree of clustering of seismic events. A change in fractal dimension corresponds to the dynamic evolution of the states of the system. Generally the values of fractal dimension cluster around 1, when the system is relatively stable and decreases to lower values around 0.3 prior to the failure. The decrease in value of fractal dimension before a big earthquake is observed by several authors (Ouchi and Uekawa 1986, De Rubies et al. 1993).

Hirata et al. (1987) showed that the change in fractal dimension with time is largely dependent on the crustal conditions of the study region. In the case of constant differential stress experiment, fractal dimension decreases along with the evolution of the fracture process. On the other hand, fractal dimension increases when differential stress increases at a constant rate. These changes are also reflected in the generalized dimension spectra. Hence, the importance of fractal dimensional analysis, as a robust index to follow the evolution of spatial and temporal distribution of seismicity in relation to the occurrence of large earthquakes has been emphasized (Dimri 2000).

1.4.4 Concept of self-organized criticality

The concept of self organized criticality (SOC) was first introduced by Bak et al. (1987; 1988), which is defined as a natural system in a marginal stable state, when the system is perturbed from the state of marginal criticality it evolves back into it. An example of self-organized critical system is a sand pile, where sand is slowly dropped onto a surface, forming a pile. As the pile grows, avalanches occur, which carry sand from the top to the bottom of the pile. In model systems, the slope of the pile becomes independent of the rate at which the system is driven by dropping sand. This is a self-organized critical slope. This concept of SOC has been widely used to explain the existence of self-similar processes like crustal deformation etc.

In seismology, the most commonly cited example of SOC is the Gutenberg-Richter magnitude frequency relationship. Earthquakes are largely responsible for the spatio-temporal fluctuations in strain hence, correlation between the scaling exponents of the spatial, temporal and magnitude distribution of earthquakes may be ascribed to the self-organization of crustal deformation. According to Hanken (1983), the macroscopic properties of self-organized systems may change systematically with time due to the perturbations in physical state of system, while the whole crust would be in a critical state and the mechanism of earthquake generation can be self organized. A good example of SOC is series of earthquakes in Koyna, India (Mandal et al. 2005).

Analysis of Himalayan earthquakes like Chamoli, 1999 and Uttarkashi, 1991 of India reveal multifractal behavior, which mainly depends on physical state of the system; hence there exists a direct relation of fractal behavior with seismotectonics of the region. The multifractal behavior of both the earthquakes can be linked with multiphased process of active Himalayan tectonics, which is a frontier research problem. However the aftershock sequence of 1993 Latur earthquake that occurred in stable continent region of India exhibited monofractal behaviour (Ravi Prakash and Dimri 2000).

1.4.5 Earthquake forecasting

In order to understand the mechanism of earthquake occurrence, more
seismic stations, monitoring micro seismicity of the region are required.
These data sets need to be analyzed in proposing predictive model. The
earthquakes are notoriously difficult to predict because the underlying mechanics that produces them is likely to be chaotic. The chaotic behavior of
slider-block models is strong evidence that the behavior of the Earth's
crust is also chaotic. This implies that exact prediction of the earthquake is
not possible, but it does not imply that earthquakes cannot be forecasted
with considerable accuracy. Quake forecasting declares that a certain
tremor has a certain probability of occurring within a given time, not that
one will definitely strike. Fractals, defined previously as mathematical
formula of a pattern that repeats over a wide range, sizes, and time scales
are hidden within the complex Earth system. By understanding the fractal
order and scale embedded in pattern of chaos, we can obtain a deeper understanding that can be utilized in forecasting the earthquake events. Although the technique is still maturing, it is expected to be reliable enough
to make official warnings possible in future.

Research is being carried out to unify scaling laws like Gutenberg–
Richter law, the Omori law of aftershocks, and the fractal dimensions of
the faults that provides a framework for viewing the probability of the occurrence of earthquakes in a given region and for a given magnitude. These
laws can be unified together by a single scaling relation (Christensen et al.
2002), which can be turned into an equation that can be solved to find the
probability that one or more aftershocks in a given magnitude range will
occur within a specified time range. This will assist seismologists to fore-
casts for aftershock sequence activities.

Fractals and forecasting of tsunami

The North Sumatra earthquake (M=9.3) of December 26, 2004 spawned a
gigantic tsunami in the Indian Ocean that completely washed out the lives
and property in coastal areas of many south-east Asian countries (like Indonesia, Srilanka, Thailand and India), has drawn serious attention of the
seismologists (Gupta 2005, Raju et al. 2005).

Tsunami, a Japanese word which means harbour wave is defined as an
ocean wave of local or distant origin that results from large-scale seafloor
displacements associated with large earthquakes, major submarine slides,
or exploding volcanic islands. These waves are capable of causing considerable destruction in coastal areas, especially where underwater earthquakes occur. Tsunami wave travels at the speed proportional to $\sqrt{gh}$,

where 'g' is acceleration due to gravity and 'h' is the depth of water column. For example tsunami wave takes a speed of about 720 Km/hr to 360 Km/ hr for ocean depth varying from 4Km to 1Km respectively. As soon as the wave reaches the shore, its speed reduces but following the principle of conservation of energy, the amplitude increases and the run-up occurs. Hence, in the open sea the wave height may be less than 1 m, it steepens to height of 15 m or more in shallow water to cause severe damage. There exists a lead time before the arrival of tsunami after an underwater displacement, which leaves a room for an effective tsunami warning system to play an important role in hazard reduction.

Better understanding of tsunami generation process is also important for numerical modeling of tsunami (Kowalik and Murty 1993 a, b). The finite difference and finite element methods have been widely used to model tsunami waves. Numerical models for different ocean domains, such as open ocean propagation or near coast propagation requires semitransparent boundary conditions to connect high resolution calculation to propagation models (Kowalik 2003). Tsunami wave propagation modeling in Indian ocean is different from the Pacific ocean. In order to map irregular coastlines, Dimri (2005), stressed the need of fractals in modeling of Tsunami for estimation of hazard.

The tsunami prediction studies (eg. Choi et al. 2002, Mofjeld et al. 1999, Abe 1995, Pelinovsky 1989) involve scaling relationships to describe the tsunami run-up heights, which can provide a basis for probabilistic forecasting of size and number of these devastating events. The power-law scaling is observed for the cumulative frequency-size distributions for tsunami run-ups recorded in Japan (Burroughs and Tebbens 2005). Authors of this paper claim that the scaling relationship can be used for probabilistic forecasting of the recurrence interval of future tsunami events within the range of run-up heights observed. Such studies to predict the recurrence intervals of large tsunami events can help in reduction of severe damage in future.

1.5 Wavelet transform: a new tool to analyze fractal signals

A wavelet is defined as a function that integrates to zero and oscillates. Wavelet transform as introduced by Grossmann and Morlet (1984) is a filter whose effective width is generally increased by powers of two (Malamud and Turcotte 1999). When this filter is passed over a time series it gives information corresponding to all scales, which helps in quick detection of noise components. The wavelet analysis involves a series of high and low pass filters convolved with the input signal, where the high pass filter is the wavelet function and the low pass filter is its scaling function. Multiresolution analysis of any data set using wavelet transform gives broad information with the scaling function and the detailed information with the wavelet function. This segregates small scaled features from large-scaled ones and helps in quick identification of different components present in the signal. The wavelet transform has got a fractal basis and is being widely used for analysis of non-stationary time series.

1.5.1 Wavelet variance analysis

The wavelet transform is a filter whose effective width is generally increased by powers of two. The wavelet transform of any time series f (t) can be written as

$$W(x, y) = \frac{1}{\sqrt{|x|}} \int_{-\infty}^{\infty} f(t) \Psi \left(\frac{t-y}{x} \right) dt \qquad (1.7)$$

where $\Psi \left(\frac{t-y}{x} \right)$ is a wavelet, W (x, y) are wavelet coefficients generated as a function of x and y, 't' is time, 'x' is time scaling or dilation and, 'y' is time shift or translation. Small values of scale parameter 'x' represent high frequency components of the signal; whereas large values of 'x' represent low frequency components of the signal. Thus the wavelet analysis of a time series at smaller scales represents detailed view and analysis at larger scales represent broader view of the data. Computation of wavelet absolute coefficients at various scales by changing the value of 'x' is the basis of multi-scale analysis.

The variance of wavelet coefficients obtained at various scales is found to follow a power law relation with scale 'x' for aftershocks sequence of 2001 Bhuj earthquake India (Fig.1.5). Mathematically this relation can be written as

$$V_w \sim x^{\alpha} \qquad (1.8)$$

where V_w is the variance of wavelet coefficients obtained at various scales and α is the wavelet exponent known as Hölder exponent. This exponent can be used to compute the fractal dimension also. The Wavelet transform with fractal basis, is becoming an indispensable signal and image processing tool for a variety of geophysical applications.

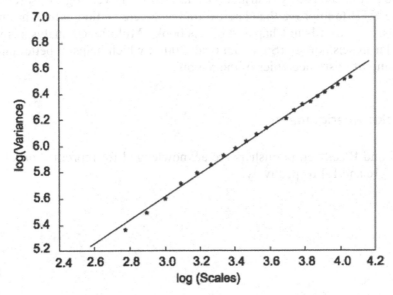

Fig. 1.5 Wavelet variance analysis at small scales of aftershocks of 2001Bhuj earthquake of India (after Dimri et al 2005)

1.6 Discussion

In previous sections, role of fractals in potential field exploration method and in understanding Earth's non-linear processes like earthquakes has been emphasized. The theory and applications of fractals is developing very rapidly. Over the past decade, many conventional techniques like Fourier analysis are being complimented by advanced techniques like wavelet analysis.

As discussed above, if a single power-law exponent is sufficient to discuss the statistics of a data set, we term it as a monofractal case otherwise while dealing with most of the geophysical data sets we talk about multifractal models. The term multifractal was first introduced by Frisch and Parisi (1985) and in view of certain limitations it became necessary to go from fractal sets to multifractal measures (Mandelbrot 1989). The main property of multifractals is the infinite hierarchy of statistical exponents that offer a very convenient framework to quantify the complex Earth system.

The notion that fractals and multifractals are relevant in geophysical studies leads to revision of the methods of geo-exploration (Turcotte 1989). Many authors like Fedi (2003) have carried out multifractal analysis

of the physical quantity distributions, derived from well log measurements
of the KTB to improve the classical rock characterization, which has been
discussed in details in Chapter 4 of this book. Multifractal analysis is very
popular in seismology (Sunmonu et al. 2001), which helps in better understanding of seismotectonics of the region.

Acknowledgements

AEG and Blackwell publishers are acknowledged for reproduction of figures 1.1c and 1.4 respectively.

1.7 References

Abe K (1995) Modelling of runup heights of Hokkaido-Nansei-Oki tsunami of 12July 1993. Pure Appl Geophys 145:735-745

Bak P, Tang C, Wiesenfeld K (1988) Self-organized criticality. Phys Rev A 38: 364-374

Bak P, Tang C, Wiesenfeld K (1987) Self-organized criticality: An explanation of 1/f noise. Phys Rev Lett 59: 381-384

Bansal AR, Dimri VP (1999) Gravity evidence for mid crustal structure below Delhi fold belt and Bhilwara super group of western India. Geophys Res Lett 26: 2793-2795

Bansal AR, Dimri VP (2001) Depth estimation from the scaling power spectral density of nonstationary gravity profile. Pure Appl Geophys 158: 799-812

Burrough MS, Tebbens SF (2005) Power-law scaling and probalistic forecasting of tsunami runup heights. Pure Appl Geophys 162:331-342

Choi BH, Pelinovsky E, Ryabov I, Hong SJ (2002) Distribution functions of Tsunami wave heights. Natural Hazards 25:1-21

Christensen K, Danon L, Scanlon T, Bak P (2002) Unified scaling law for earthquakes. Proc Natl Acad Sci USA 99:2509-2513

Constable SC, Parker RL, Constable CG (1987) Occam's inversion: A practical algorithm for generating smooth models from electromagnetic sounding data. Geophysics 52: 289-300

De Rubies V, Dimitriu P, Papa Dimtriu E, Tosi P (1993) Recurrent patterns in the spatial behavior of Italian seismicity revealed by the fractal approach. Geophys Res Lett 20:1911–1914

Dimri VP (1992) Deconvolution and inverse theory, Elsevier Science Publishers, Amsterdam London New York Tokyo

Dimri VP (1998) Fractal behavior and detectibility limits of geophysical surveys. Geophysics 63:1943–1946

Dimri VP (2000) Application of fractals in earth science. AA Balkema, USA and Oxford IBH Publishing Co New Delhi

Dimri VP (2005) Tsunami wave modeling using fractals and finite element technique for irregular coastline and uneven bathymetries. Natl workshop on formulation of science plan for coastal hazard preparedness (18-19 Feb) NIO, Goa, India

Dimri VP, Nimisha V, Chattopadhyay S (2005) Fractal analysis of aftershock sequence of Bhuj earthquake - a wavelet based approach. Curr Sci (in press)

Fedi M (2003) Global and local multiscale analysis of magnetic susceptibility data. Pure Appl Geophys 160:2399–2417

Fedi M, Quarta T, Santis AD (1997) Inherent power-law behavior of magnetic field power spectra from a Spector and Grant ensemble. Geophysics 62:1143-1150

Frisch U, Parisi G (1985) Fully developed turbulence and intermittency. In: Ghill M (ed) Turbulence and predictability in geophysical fluid dynamics and climate dynamics, North Holland, Amsterdam, pp 84

Grant FS (1985) Aeromagnetics, geology and ore environments 1. Magnetite in igneous, sedimentary and metamorphic rocks: An overview. Geoexploration 23:303-333

Gregotski ME, Jensen OG, Arkani-Hamed J (1991) Fractal stochastic modeling of aeromagnetic data. Geophysics 56:1706-1715

Grossmann A, Morlet J (1984) Decomposition of Hardy functions into square integrable wavelets of constant shape. SIAM J Math Anal 15:723-736

Gupta HK (2005) A Note on the 26 December 2004 tsunami in the Indian ocean. J Geol Soc India 65: 247-248

Hanken H (1983) Advanced synergetics: Instability hierarchies of self-organizing systems and devices, Springer, Berlin Heidelberg New York

Hirabayashi T, Ito K, Yoshi (1992) Multifractal analysis of earthquakes. Pure Appl Geophys 138: 591-610

Hirata T, Imoto M (1991) Multifractal analysis of spatial distribution of micro earthquakes in the Kanto region. Geophys J Int 107:155–162

Hirata T, Sato T, Ito K (1987) Fractal structure of spatial distribution of microfracturing in rock. Geophys J R A Soc 90:369-374

Kagan YY, Knopoff L(1980) Spatial distribution of earthquakes. The two point correlation function. Geophys J R A Soc 62:303–320

Kanamori H, Anderson DL(1975) Theoretical basis of some empirical relations in seismology. Bull Seis Soc America 65:1073-1095

Korvin G (1992) Fractal models in the earth sciences. Elsevier Science Publishers, Amsterdam London New York Tokyo

Kowalik Z (2003) Basic relations between tsunamis calculation and their physics–II. Sci Tsunami Haz 21: 154-173

Kowalik Z, Murty TS (1993a) Numerical modeling of ocean dynamics. World Sci Publ, Singapore New Jersey London Hong Kong

Kowalik Z, Murty TS (1993b) Numerical simulation of two-dimensional tsunami runup. Marine Geodesy 16: 87–100

Lovejoy S, Schertzer S, Ladoy P (1986) Fractal characterization of homogeneous geophysical measuring network. Nature 319:43-44

Malamud BD, Turcotte DL (1999) Self affine time series I: generation and analysis. In: Dmowska R, Saltzman B (ed.) Advances in Geophysics: Long Range Persistence in Geophysical Time Series, vol 40. Academic Press, San Diego, pp 1-87

Mandal P, Mabawonku AO, Dimri VP (2005) Self-organized fractal seismicity of reservoir triggered earthquakes in the Koyna-Warna seismic zone, Western India. Pure Appl Geophys 162: 73-90

Mandelbrot BB (1983) The Fractal Geometry of Nature. WH Freeman & Company, New York

Mandelbrot BB (1989) Mulifractal measures, especially for geophysicists. In: Scholz CH, Mandelbrot BB(eds) Fractals in geology and geophysics, Berkhäuser Verlag, Basel, pp 5-42

Mandelbrot BB, Van Ness JW (1968) Fractional Brownian motions, fractional noises and applications. SIAM rev.10:422-437

Maus S (1999) Variogram analysis of magnetic and gravity data. Geophysics 64:776-784

Maus S, Dimri VP (1994) Fractal properties of potential fields caused by fractal sources. Geophys Res Lett 21: 891-894

Maus S, Dimri VP (1995) Potential field power spectrum inversion for scaling geology. J Geophys Res 100: 12605–12616

Maus S, Dimri V (1996) Depth estimation from the scaling power spectrum of potential fields? Geophys J Int 124: 113-120

Maus S, Gordon D, Fairhead JD(1997) Curie-depth estimation using a self-similar magnetization model. Geophys J Int 129:163-168

Mofjeld HO, Gonzalez FI, Newman JC (1999) Tsunami prediction in coastal regions. In: Mooers CNK (ed) Coastal ocean prediction, Am Geohys Union, Washington DC, pp 353-375

Moharir PS (2000) Multifractals. In: Dimri VP (ed) Application of fractals in earth sciences AA Balkema, USA Oxford IBH Pub Co New Delhi, pp 46-57

Naidu P (1968) Spectrum of the potential field due to randomly distributed sources, Geophysics 33: 337-345

Omori (1894) On aftershocks. Rep Imp Earthq Investig Comm 2:103-139 (in Japanese)

Ouchi T, Uekawa T (1986) Statistical analysis of the spatial distribution of earthquakes before and after large earthquakes. Phys Earth Planet Int 44: 211-225

Pelinovsky EN (1989) Tsunami climbing a beach and Tsunami zonation. Sci Tsunami Haz 7:117-123

Peterson I (1984) Ants in the Labyrinth and other fractal excursions. Science News 125: 42- 43

Pilkington M (1997) 3-D magnetic imaging using conjugate gradients. Geophysics 62:1132-1142

Pilkington M, Todoeschuck JP (1990) Stochastic inversion for scaling geology. Geophys J Int 102:205-217

Pilkington M, Todoeschuck JP (1991) Naturally smooth inversions with a priori information from well logs. Geophysics 56:1811-1818

Pilkington M, Todoeschuck JP (1993) Fractal magnetization of continental crust. Geophys Res Lett 20:627-630

Pilkington M, Todoeschuck JP (1995) Scaling nature of crustal susceptibilities. Geophys Res Lett 22:779-782

Pilkington M, Todoeschuck JP (2004) Power-law scaling behavior of crustal density and gravity. Geophys Res Lett 31: L09606, doi: 10.1029/2004GL019883

Pilkington M, Gregotski ME, Todoeschuck JP (1994) Using fractal crustal magnetization models in magnetic interpretation. Geophys Prosp 42:677-692

Quarta T, Fedi M, Santis AD (2000) Source ambiguity from an estimation of the scaling exponent of potential field power spectra. Geophys J Int 140: 311-323

Raju PS, Raghavan RV, Umadevi E, Shashidar D, Sarma ANS, Gurunath D, Satyanarayana HVS, Rao TS, Naik RTB, Rao NPC, Kamuruddin Md, Gowri Shankar U, Gogi NK, Baruah BC, Bora NK Kousalya M, Sekhar M, Dimri

VP(2005) A note on 26 December,2004 Great Sumatra Earthquake. J Geol Soc India 65: 249-251

Ravi Prakash M, Dimri VP (2000) Distribution of the aftershock sequence of Latur earthquake in time and space by fractal approach. J Geol Soc India 55: 167-174

Saggaf M, Toksoz M (1999) An analysis of deconvolution: Modeling reflectivity by fractionally integrated noise. Geophysics 64:1093-1107

Spector A, Grant FS (1970) Statistical models for interpreting magnetic data. Geophysics 35:293- 302

Sunmonu LA, Dimri VP, Ravi Prakash M, Bansal AR (2001), Multifractal approach of the time series of $M \geq 7$ earthquakes in Himalayan region and its vicinity during 1985-1995. J Geol Soc India 58: 163-169

Thorarinsson, F, Magnusson SG (1990) Bouguer density determination by fractal analysis. Geophysics 55: 932–935

Todoeschuck JP, Jensen OG (1988) Joseph geology and scaling deconvolution. Geophysics 53:1410-1411

Todoeschuck JP, Jensen OG (1989) Scaling geology and seismic deconvolution. Pure Appl. Geophys 131: 273-288

Toverud T, Dimri VP, Ursin B (2001) Comparison of deconvolution methods for scaling reflectivity. Jour Geophy 22:117-123

Tubman KM, Crane SD (1995) Vertical versus horizontal well log variability and application to fractal reservoir modeling. In: Barton CC, LaPointe PR (eds) Fractals in Petroleum Geology and Earth Sciences, Plenum press, New York, pp 279-294

Turcotte DL (1989) Fractals in geology and geophysics. In: Scholz CH, Mandelbrot BB (eds) Frcatals in Geophysics, Berkhäuser Verlag, Basel, pp 171-196

Turcotte DL (1992) Fractals and chaos in geology and geophysics. Cambridge University Press, Cambridge

Chapter 2. Fractal Modeling of Complex Subsurface Geological Structures

V. P. Dimri, Ravi P. Srivastava

National Geophysical Research Institute, Hyderabad, India

2.1 Summary

The essential component of gravity modeling is an initial model with arbitrary shape having regular geometry. This regular geometry approximates causative body of irregular geometry. For best approximation of causative bodies using regular geometry one requires several polygons represented by many vertices, which are perturbed during global optimization to achieve best model that fits the anomaly. We have circumvented the choice of multi-face regular polygonal initial model by using L_p norm modified Voronoi tessellation. This tessellation scheme provides realistic irregular (fractal) geometry of the causative body using a few parameters known as Voronoi centers, which makes inversion algorithm faster as well as provides an irregular realistic final model for the causative body.

2.2 Introduction

There are two ways to interpret the gravity data viz. forward and inverse. According to the equivalent layer theorem there are infinite models whose theoretical response could fit the observed gravity data, but the number of possible models is reduced if a priory guess about the source model is known (Dimri 1992). Depending on the non-linearity relation between the change in data parameter and its response to model parameter, Dimri (1992) grouped the inversion scheme in following four categories:
i) linear ii) weakly non-linear iii) quasi non-linear and iv) highly non-linear. The methods like generalized inversion, singular value decomposition, gradient methods, Monte Carlo method and its variants like simulated annealing and genetic algorithms are recommended depending on non-linearity in the system, but success of all these methods depends on the accuracy of the forward modeling. Hence in this chapter a new approach ap-

plying fractals is used to formulate the problem suitable for the automated inversion schemes.

In case of geophysical studies the density contrast between different interfaces is responsible for the gravity anomaly. Nettleton (1940;1942) used two simple methods for making approximate gravity anomaly calculations. One involves the use of circular discs and the other uses an end correction to two-dimensional bodies. Bott (1960) suggested a method to trace the floor of a sedimentary basin, which involved the approximation of a sedimentary basin by a series of two dimensional juxtaposed rectangular/square blocks of uniform density. Computation of gravity anomaly due to 2D and 3D bodies of arbitrary shape (Talwani et al. 1959, Talwani and Ewing 1960) uses number of contours to approximate the body and these contours are replaced by polygonal lamina for computation of the anomaly. Several other workers pioneered different approaches of gravity anomaly computation viz. computation of gravity response using, mechanical integrator (Arnold 1942), fast Fourier transform (Parker 1973, Bhattacharya and Navolio 1976) and complex polynomials (Sergio and Claudia 1997) but all of them have used regular geometry of the causative body as an initial model.

Generally many patterns of nature are so irregular and fragmented, that, compared with Euclid geometry nature exhibits not simply a higher degree but an altogether different level of complexity (Mandelbrot 1983). In geophysical context, almost all the natural bodies under study are highly irregular. Existence of these patterns entails to study those forms that Euclid leaves aside as being formless to investigate the causative body.

In the fractal models of the porous rocks (Turcotte 1997, Korvin 1992), there is a tendency to focus on the rock matrix to predict physical properties, viz. density, susceptibility; useful in mineral exploration. Another aspect of fractals for porous rocks is the distribution and connectivity of the pore space to predict transport properties, and percolation clusters; useful in petroleum and groundwater study. The similar property has been used for earthquakes study in chapter 5.

Dimri (2000) used the concept of fractals for studying the flow media and also opined scaling behavior of potential fields (Gregotski et al. 1991, Pilkington and Todoeschuck 1993, Maus and Dimri 1994; 1995; 1996, Dimri 2000) by establishing a relation between the scaling exponent of source and field, which is useful for understanding of fractal geology.

The problem we shall take up is the following: Given a gravity anomaly along a profile or over an area, what is the shape of a 2D/3D causative body of varying physical property, which will produce this anomaly. Recent and ongoing attempts have centered about solving the problem by automated inversion of assumed polygonal causative body either by perturbation scheme (Bott 1960, Corbato 1965, Tanner 1967, Negi and Garde 1969) or by using iterative procedure of fast Fourier transform method for calculating the shape of perturbing body (Parker 1973, Oldenburg 1974), which leads to numerous disadvantages (Oldenburg 1974).

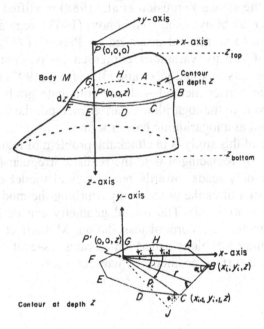

Fig. 2.1 Approximation of arbitrary 3D body (after Talwani and Ewing 1960)

2.3 Conventional forward method

In conventional methods the forward problem involves computation of gravity effects of geological body described by a closed polygon as shown in Fig. 2.1 (Talwani and Ewing 1960). The computed gravity value is compared with the observed gravity value at each point and the difference is given by $g_{diff} = g_{cal} - g_{obs}$.

In each iteration depth/density contrast of the source is altered and new gravity response is computed until the difference between computed and observed is minimum as decided by the interpreter (Dimri 1992). This technique is called trial and error and has been used by Talwani and Heirtzler (1964) and many others and even now the modified version of this technique imposed with certain constraints is used for the inverse modeling of gravity data. Inverse modeling is also done in frequency domain. Oldenburg (1974) has slightly modified the fast computation of gravity anomaly using fast Fourier transform proposed by Parker (1973) for inverting the gravity data and later Li and Oldenburg (1997) used wavelets for fast inversion of the magnetic data. These authors have exploited data compression property of the wavelets and used sparse matrices to reconstruct the signal. Fergusion et al. (1988) modified the method and achieved better stability using Tikhonov (1963) regularization. Guspi (1992) used the same method proposed by Parker (1973) with the slight modification of density variations expressed by polynomials in z with variable coefficients in x and y. Li and Oldenburg (1997) have used a primal logarithmic barrier method with the conjugate gradient technique as the central solver. In the logarithmic barrier method, the bound constraints are implemented as a logarithmic barrier term.

The purpose of this study is to attack the problem of regular geometry of causative body by replacing it with the realistic irregular (fractal) geometry, which not only leads towards realistic final model of the causative body but also simplifies the process of perturbing the model during global optimization (Dimri 1992). The fractal geometry can be generated using very few parameters by Voronoi tessellation. Moharir et al. (1999) have also used a similar technique called lemniscates tessellation for global optimization of sub-surface geometry by hamming scan.

2.4 Theory

There are a variety of algorithms available to construct Voronoi diagrams (Lee 1982, Okabe et al. 1992). One popular method known as sweep line algorithm is the incremental algorithm that adds a new site to an already existing diagram (Fortune 1987).

Given a set S of n distinct points in R^d, Voronoi diagram is the partition of R^d into n polyhedral regions V(p). Each region V(p), called the Voronoi cell of point 'p' is defined as the set of points in R^d which are closer to 'p' than any other arbitrary point 'q' in S, or more precisely,

$$V(p) = \left\{ x \in R^d \ \text{dist}(x,p) \leq \text{dist}(x,q) \, \forall q \in S - p \right\} \tag{2.1}$$

where, 'dist' is the Euclidean distance function.

In a straight forward iterative algorithm for the planar Voronoi diagram Tipper (1990) illustrated that the Voronoi tessellation in a two dimensional space consists of enclosing every center by a Voronoi polygon (Fig.2.2) such that common edge of adjacent polygons is perpendicular bisector to the line joining the centers on each side of that edge. Here we have generalised the notion of Voronoi tessellation by using L_P distances instead of the Euclidian distances, so that Voronoi domains are not necessarily of polygonal shape. The L_p distance is given by the expression:

$$L_p = (x - p_j)^{1/p} \tag{2.2}$$

where x is an arbitrary point and P_j is a vector whose distance has to be calculated, and p is an exponent which can hold any real value, j=1:N, where N is the number of Voronoi centers.

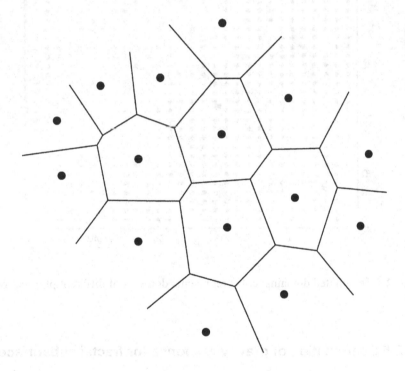

Fig. 2.2 Voronoi polygons corresponding to the Voronoi centers shown as black dots

Initially a few Voronoi centers are taken and from them two dimensional tessellated region is generated in which, domains with different physical properties are shown in different colours (Fig. 2.3). The present geometrical representation brings a new facet of domain characterization by a set of parameters, referred herein as Voronoi centers. These parameters are perturbed and thus the different tessellated regions are generated at different depths. This characterization method entails the development for the solution of geophysical inverse problems with the help of global optimization techniques. Assigning density contrast to regions of interest is accomplished during the tessellation of domains using modified Voronoi tessellation method.

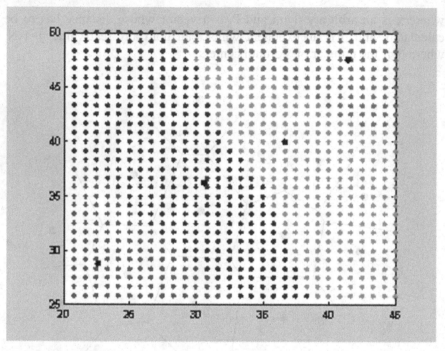

Fig. 2.3 Tessellated domains: color represents domains of different physical properties

2.5 Computation of gravity response for fractal subsurface

The forward gravity response due to each tessellated region at depth Z is calculated at each node of the grid laid at the surface. The computation of gravity response is repeated for another tessellated regions at different depths, and then integrated response is calculated at each node of the surface grid (Fig. 2.4) by numerical integration with depth. If tessellated regions are at equal depths then the integrated response can be calculated by Simpson's rule otherwise Gauss's quadrature formula can be used for computation of the cumulative response due to tessellated regions at unequal depths.

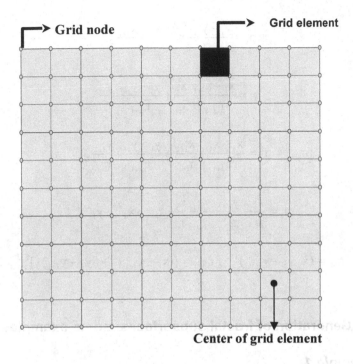

Center of grid element

Fig. 2.4 Grid laid over the area of interest on the surface

2.5.1 Mathematical expressions for computation of the gravity anomaly

The gravity anomaly caused by the polygonal lamina per unit thickness, in a form suitable for programming is expressed in terms of the co-ordinates of two successive vertices of the polygon viz. x_i,y_i,z and x_{i+1},y_{i+1},z, as given by (Talwani and Ewing 1960)

$$V = G\rho \sum_{i=1}^{n} [W \arccos\{(x_i/r_i)(x_{i+1}/r_{i+1})+(y_i/r_i)(y_{i+1}/r_{i+1})\}$$

$$-\arcsin\frac{zq_iS}{(p_i^2+z_i^2)^{1/2}}+\arcsin\frac{zf_iS}{(p_i^2+z_i^2)^{1/2}}]$$

where S= +1 if p_i is positive, S= -1 if p_i is negative, W= +1 if m_i is positive, W= -1 if m_i is negative, 'Z' is depth and 'n' is number of sides in the polygon, and

$$p_i = \frac{y_i - y_{i+1}}{r_{i,i+1}} x_i - \frac{x_i - x_{i+1}}{r_{i,i+1}} y_i,$$

$$q_i = \frac{x_i - x_{i+1}}{r_{i,i+1}} \frac{x_i}{r_i} + \frac{y_i - y_{i+1}}{r_{i,i+1}} \frac{y_i}{r_i},$$

$$f_i = \frac{x_i - x_{i+1}}{r_{i,i+1}} \frac{x_{i+1}}{r_{i+1}} + \frac{y_i - y_{i+1}}{r_{i,i+1}} \frac{y_{i+1}}{r_{i+1}},$$

$$m_i = \frac{x_{i+1}}{r_{i+1}} \frac{y_i}{r_i} - \frac{y_{i+1}}{r_{i+1}} \frac{x_i}{r_i}, \; r_i = (x_i^2 + y_i^2)^{1/2},$$

$$r_{i+1} = (x_{i+1}^2 + y_{i+1}^2)^{1/2}, \; r_{i,i+1} = [(x_i - x_{i+1})^2 + (y_i - y_{i+1})^2]^{1/2},$$

2.6 Generation of fractal subsurfaces: some examples

Example 1

A fractal subsurface is generated using L_p norm taking p=1.5, wherein the Voronoi region is defined by the co-ordinates x= 20 to 45 and y= 25 to 50. The Voronoi centers chosen within this region are given by the following co-ordinates:

x	y
22.6	28.75
30.6	30.6
36.6	36.6
41.4	41.4

The subsurface thus generated is shown in Fig 2.3 with Voronoi centers marked as black dots.

Example 2

In another example the fractal subsurface at different depth levels are generated using different L_p norms. The coordinates of Voronoi region for all the subsurfaces were taken as x = 20 to 50 and y = 5 to 65 and Voronoi centers within the region were taken as:

x	y
22.0	26.0
30.0	33.0
37.0	40.0
46.0	48.0
48.0	58.0

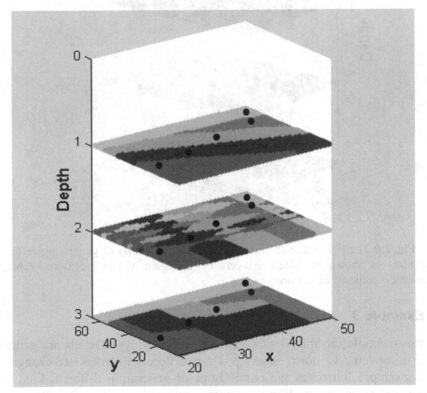

Fig. 2.5 The 3-D subsurface structure wherein 2-D layers of variable physical property regions are overlain to generate 3-D volume. The figure demonstrates the ability of generating various kind of structures shown in different layers merely by changing exponent p in L_p norm, keeping Voronoi centers fixed

This example shows the variation in geometry and provides an excellent way of changing the geometry merely by changing the exponent p in L_p norm. The results are shown in Fig. 2.5 where the first layer (topmost) corresponds to p=1.5, the middle layer corresponds to p=-1.5 and the lowermost corresponds to p=1.0, which is equivalent to L_1 norm.

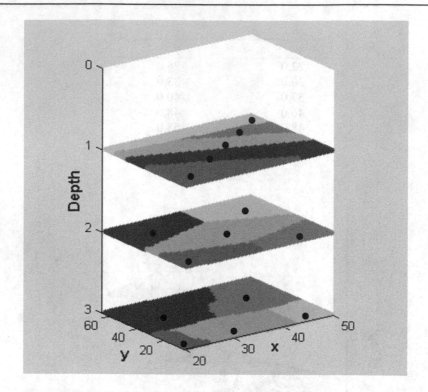

Fig. 2.6 The figure demonstrates the another possibility of generating various structures as shown in different layers by changing Voronoi centers keeping exponent p constant in L_p norm

Example 3

In this example the fractal subsurface at different depth levels are generated using same L_p norm, where p=1.5 but Voronoi centers are changed. This example illustrates the possibilities of generating different kind of structures by changing Voronoi centers as shown in Fig. 2.6. The co-ordinates of Voronoi region for all the subsurfaces were taken as x= 20 to 50 and y= 5 to 65 and 5 Voronoi centers within the Voronoi region were taken with following x, y co-ordinates:

Top layer		Middle Layer		Bottom Layer	
x_1	y_1	x_2	y_2	x_3	y_3
22.0	26.0	22.0	7.0	15.0	8.0
30.0	33.0	22.0	50.0	30.0	50.0
37.0	40.0	35.0	33.0	35.0	10.0
46.0	48.0	50.0	7.0	45.0	45.0
48.0	58.0	50.0	50.0	55.0	10.0

Example 4

This example shows the gravity anomaly computed over the fractal subsurface is shown (Fig. 2.7). Fig. 2.7(a) corresponds to the particular case wherein Voronoi centers are taken along a line which gives layered model where as Fig. 2.7(b) represents the tessellated fractal structure bounded by x=20 to 50 and y= 5 to 65, wherein the Voronoi centers correspond to those given in Example 1. The density values corresponding to red, blue, green and magenta colors are 2.1, 2.3, 2.67 and 2.5 respectively. The fractal subsurface is assumed at the depth of 10 unit from the surface.

```
%VOR.M: program for generating fractal structures by Voronoi tessellation us-
ing L(p) distances


hp=[ ];
reg=[10 25 10 25];

p=2;
n=4;
a=reg(1);b=reg(2);c=reg(3);d=reg(4);
colour=['.r' '.b' '.g' '.m' '.c' '.k' '.y' '.w'];
xc=a+(b-a)*rand(1,n);
zc=c+(d-c)*rand(1,n);

figure(1)
xdiff=b-a; zdiff=d-c;
np=50;
for i = 1:np
for j = 1:np
x=a+(b-a)*i/np;
z=c+(d-c)*j/np;
dp=((abs(x-xc)).^p + (abs(z-zc)).^p).^(1/p);
[dpmin,k(i,j)]=min(dp);
for ni=1:n
if (k(i,j)==ni)
   plot(x,z,colour(2*ni-1:2*ni),'MarkerSize',10);
   hp=[hp;x,z,ni,dp];
axis(reg)
hold on
end
end
end
end
hold off
```

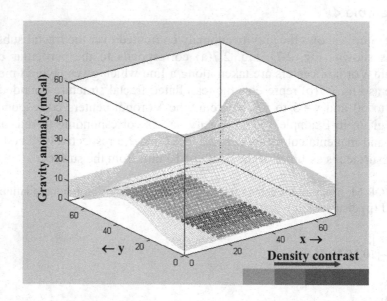

Fig. 2.7 (a) Gravity anomaly response over the simplified horizontal layered model

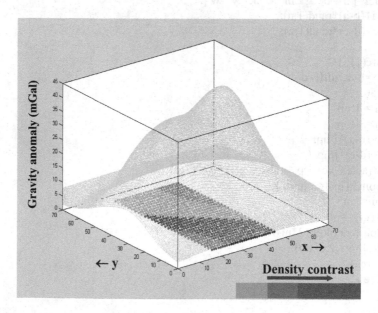

Fig. 2.7 (b) Gravity response calculated over fractal subsurface structure showing four regions of different physical property (density) variations

2.7 Discussions

The new method to generate fractal geometry using modified Voronoi tessellation discussed here is very useful for any kind of inverse and forward geophysical modeling. The geometry of natural sources is so complex that approximating it with regular geometry may not be useful in case of high resolution studies. In case of potential fields it may be bypassed because of low resolution of the potential field data but the given technique has another advantage of using very few parameters, hence it provides faster solutions than the conventional methods wherein for describing geometry itself many vertices are involved.

Another application of this method is in understanding the movement of oil front for reservoir monitoring programmes using gravity measurements.

Acknowledgement

Authors are thankful to SEG publishers for their permission to reproduce figure 2 1.

2.8 References

Arnold JF, Siegert (1942) A mechanical integrator for the computation of gravity anomalies. Geophysics 74:354-366

Bhattacharya BK , Navolio ME (1976) A fast Fourier transform method for rapid computation of gravity and magnetic anomalies due to arbitrary bodies, Geophys Prosp 20: 633-649

Bott MHP (1960) The use of rapid digital computing methods for direct gravity interpretation of sedimentary basins. GJRAS 3:63-67

Carbato CE (1965) A least squares procedure for gravity interpretation. Geophysics 30:228-233

Dimri VP (1992) Deconvolution and inverse theory. Elsevier Science Publishers, Amsterdam London New York Tokyo

Dimri VP (2000) Fractal dimension analysis of soil for flow studies. In: Application of fractals in earth sciences, edited, V.P.Dimri, pp 189-193, A.A. Balkema, USA

Dimri VP (2000) Crustal fractal magnetisation. In: Application of fractals in earth sciences, edited, V.P.Dimri, pp 89-95, A.A. Balkema, USA

Fortune S (1987) A sweepline algorithm for Voronoi diagrams. Algorithmica 2:153-174

Ferguson JF, Felch RN, Aiken CLV,Oldow JS, Dockery H (1988) Models of the Bouguer gravity and geologic structure at Yucca Flat Navada. Geophysics 53: 231-244

Gregotski ME, Jensen OG , Akrani – Hamed J (1991) Fractal stochastic modeling of aeromagnetic data. Geophysics 56:1706-1715

Gupsi F (1992) Three-dimensional Fourier gravity inversion with arbitrary gravity contrast. Geophysics.57:131-135

Korvin G (1992) Fractal models in the earth sciences, Elsevier Science Publishers, Amsterdam London New York Tokyo

Lee DT (1982) On k – nearest neighbours Voronoi diagrams in the plane, IEEE Transactions on Computers C-31: 478-487

Li Y, Oldenburg DW (1997) Fast inversion of large scale magnetic data using wavelets. 67th Ann. Internat. Mtg., Soc. Expl. Geophys., Expanded Abstracts, 490–493

Li Y, Oldenburg DW (1998) 3D inversion of gravity data. Geophysics 63: 109–119

Mandelbrot BB (1983) The Fractal Geometry of Nature. WH Freeman & Company, San Francisco

Maus S, Dimri VP (1994) Fractal properties of potential fields caused by fractal sources. Geophys Res Lett 21: 891-894.

Maus S, Dimri VP (1995) Potential field power spectrum inversion for scaling geology: J Geophys Res 100: 12605–12616

Maus S, Dimri VP (1996) Depth estimation from the scaling power spectrum of potential fields? Geophys J Int 124: 113-120

Moharir PS, Maru VM, Srinivas S (1999) Lemniscates representation for inversion of gravity and magnetic data through nonlocal optimization. Proc Ind Acad Sci (Earth & Planet Sci) 108:223-232

Negi JG, Garde SC (1969) Symmetric matrix method for gravity interpretation. Jour Geophys Res 74:3804-3807

Nettleton LL (1940) Geophysical prospecting for oil. McGraw Hill Book Co.

Nettleton LL (1942) Gravity and magnetic calculations. Geophysics 7:293-310

Oldenburg DW (1974) The inversion and interpretation of gravity anomalies. Geophysics 39:526-536

Okabe A, Barry B, Sugihara K (1992) Spatial tessellations: Concepts and applications of Voronoi diagrams. John Wiley & Sons

Parker RL (1973) The rapid computation of potential anomalies. Geophys J.R. Astr. Soc., 31: 447- 455

Pilkington, M., Todoeschuck, J.P.(1993) Fractal magnetization of continental crust. Geophys. Res. Lett., 20: 627-630

Sergio E. Oliva, Claudia L. Ravazzoli (1997) Complex polynomials for the computation of 2D gravity anomalies. Geophys. Pros. 45: 809

Talwani M, Worzel JL, Landisman M (1959) Rapid gravity computations for two-dimensional bodies with application to the Mendocino submarine fracture zone. J. Geophys. Res. 64: 49-59

Talwani M, Ewing M (1960) Rapid computation of gravitational attraction of three-dimensional bodies of arbitrary shape, Geophysics 25: 203-225

Talwani M, Heirtzler J.(1964) Computation of magnetic anomalies caused by 2-D structures of arbitrary shapes. In: Compt. Min. Ind. Part 1, Stanford Univ. Geo.Sci., 9

Tanner, J.G., 1967, An automated method of gravity interpretation, Geophys. J. R. Astr. Soc. 13: 339-347

Tikhonov AN (1963) Solution of incorrectly formulated problems and the regularizations method. Soviet Math. Doklady, 4:1035:1038

Tipper J.C., 1990, A straight forward iterative algorithm for the planar Voronoi diagram, Information Process Letters 34:155-160

Turcotte D., 1997, Fractals and Chaos in Geology and Geophysics, 2nd edition,: Cambridge University Press, Cambridge,New York

Chapter 3. The Route to Fractals in Magnetotelluric Exploration of the Crust

Karsten Bahr

Department of Geology and Geophysics, Adelaide University, Australia

On leave from:

Institut für Geophysik, Herzberger Landstr, Göttingen, Germany

3.1 Summary

Although electromagnetic induction is governed by a linear diffusion equation, the magnetotelluric method has provided evidence of fractal structures in the crust. During the search for an electrical conduction mechanism that is compatible with the geophysical anomalies in the middle and lower crust, random resistor networks were developed. They contain two types of resistors, representing the rock matrix and the conductive phase. Random resistor network models can explain both the electrical anisotropy and the lateral variability of the bulk conductivity found in large scale electromagnetic array experiments. These observations are a consequence of the very non-linear relationship between the amount of conductive material and the bulk conductivity of strongly heterogeneous media. Coincidence between the statistical properties of field data and modelled data is obtained if resistor networks with fractal geometry are employed. This can indicate that the natural conductive networks also have fractal geometry and stay close to a percolation threshold.

3.2 Introduction

Crustal fault networks have been studied by Gueguen, David & Gavrilenko (1991) and Gueguen, Gavrilenko and Le Ravalec (1996) in order to study the scale effect of rock permeability. Fault networks with self-similar geometry were studied in the context of earthquake cycle development by Heimpel and Olson (1996) and Heimpel (1997). This review shall provide independent evidence for the existence of these fault networks from an exploration technique which so far rarely contributed to the subject of fractals in geophysics: the magnetotelluric (MT) method. The existence of electrically conductive structures with fractal geometry emerged in two basic steps: (1) the application of a general 'anisotropy test' to MT field data shows that the well-known midcrustal and lower crustal geophysical anomalies can be electrically anisotropic (Kellet et al. 1992, Bahr et al. 2000), and (2) This anisotropy results from the anisotropic connectivity of conductive material in conductive fractures networks with fractal geometry (Bahr 1997).

With respect to the electrical conductivity of the upper mantle and transition zone, results from geophysical large scale induction studies and from laboratory studies are in reasonable agreement (Xu et al. 1998). In contrast, there is a discrepancy between the high conductivities found by EM studies focusing on the middle and lower crust (see reviews by Haak and Hutton 1986, Jones 1992) and the low conductivities found in laboratory studies of crustal rocks with appropriate temperatures and pressures (e.g. Lastovickova 1991).

The route to fractal conductors in the middle crust was to some extent paved by the search for a conduction mechanism. From the observations in the German deep drilling program, KTB it can be concluded that neither electronic conduction nor electrolytic conduction can be ruled out (ELEKTB 1997). Kontny et al. (1997) point out that even if electronic conduction is considered then fluids still play an important role in the deposition of ores and of graphite. Another result of the ELEKTB (1997) group where the distribution functions of the electrical resistivity on all scales (Fig. 3.1): while seismic velocities in the crust are observed only within a narrow band, electrical resistivity can vary over 3 orders magnitude.

Field studies by Cull (1985), Rasmussen (1988), Tezkan et al. (1992) and Kellett et al. (1992) for the first time gave evidence of electrically anisotropic structures in the crust. Interpretations ascribing conductivity anomalies in the middle or lower crust to graphite (Frost et al. 1989, Mareschal 1990, Mareschal et al. 1992, Jödicke 1992) or to brines (Shankland and Ander 1983, Gough 1986, Touret 1986, Bailey et al. 1989, Hyndman and Shearer 1989, Marquis and Hyndman 1992) must explain electrical anisotropy as an additional property. All these models incorporate a mixture of a low-conductivity rock matrix and a second highly conductive phase. Therefore the search for the conduction mechanism is related to the research on the distribution of pores and cracks (Wong et al. 1989, Allegre and LeMouel 1993) and on mixing laws (Greenberg and Brace 1969, Kirkpatrick 1973, Madden 1976, Doyen 1988).

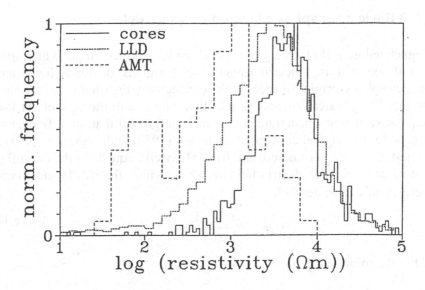

Fig. 3.1 Frequency distribution function of electrical resistivity on three scales: borehole samples, d ≈ 1 cm; laterolog, d ≈ 1m; and audiomagnetotellurics, d ≈ 1km

The model of an embedded network (Madden 1976) can explain the observed electrical anisotropy if the network is anisotropic with respect to its two different electrical connectivities in two different directions. This is an application of percolation theory (e.g. Kirkpatrick 1973, Zallen 1983, Stauffer and Aharony 1992, Gueguen and Palciauskas 1994), and those networks which exhibit strong anisotropy of their connectivity are in the vicinity of the percolation threshold (Bahr 1997).

I shall first review the basic concept of magnetotellurics and summarize the development of a series of general conductivity models anticipating the anisotropic model of Kellet et al. (1992). The discussion on the origin of the crustal conductors is mentioned only briefly here because extensive literature on this subject is available (Jones 1992, Simpson 1999; 2001, Wannamaker 2000, Yardley and Valley 1997; 2000). I finally demonstrate how random resistor networks can be used to model the conduction mechanism in very heterogeneous media. Both fractal and non-fractal networks are employed. If the conductive structure has fractal geometry then many properties of the field data and in particular the anisotropy can be explained.

3.3 Basic concept of a layered Earth model

Magnetotellurics (MT) is a geophysical exploration technique which uses natural time-varying electromagnetic (em) signals of ionospheric and magnetospheric origin for probing the conductivity distribution of the subsurface. The physical process is induction, and due to the skin effect low frequency em fields penetrate deeper into the ground than high frequency signals. In the original concept of Tikhonov (1950) and Cagniard (1953) a layered earth was considered, and from Maxwell's equations the em diffusion equation for the electric field vector E_n within the n'th layer of conductivity σ_n can be derived.

$$\nabla^2 E_n = \mu_0 \sigma_n \frac{\partial E_n}{\partial t} \tag{3.1}$$

It has the solution

$$E_{xn} = E_{x1n} e^{i\omega - q_n z} + E_{x2n} e^{i\omega + q_n z} \tag{3.2}$$

where ω is the angular frequency of the source field and

$$q_n = \frac{1}{C_n} = (i\mu_0 \sigma_n \omega)^{\frac{1}{2}} \tag{3.2a}$$

wherein C_n is a complex frequency dependent (transfer function) penetration depth within nth layer.

At the surface the transfer function C_0 or the impedance Z can be calculated from orthogonal components of the horizontal magnetic and electric field.

$$C_0 = Z / i\omega\mu_0 = \frac{Ex}{i\omega\mu_0 Hy} = -\frac{Ey}{i\omega\mu_0 Hx} \tag{3.3}$$

Wait (1954) showed that the solution of the forward problem that relates to the model parameters can be found with the recurrence formula

$$C_n(z_{n-1}) = \frac{1}{q_n} \frac{q_n C_{n+1}(z_n) + \tanh(q_n d_n)}{1 + q_n C_{n+1}(z_n) \tanh(q_n d_n)} \tag{3.4}$$

It links the transfer function C_n at the top of the nth layer with thickness $d_n = Z_n - Z_{n-1}$ to the known transfer function C_{n-1} at the bottom of the layer (Fig. 3.2). The recursion starts with the transfer function at the top of the underlying homogenous half-space with conductivity σ_n.

$$C_N = 1/q_n = (i\mu_0 \sigma_N \omega)^{-\frac{1}{2}} \tag{3.5}$$

MT data from both forward model studies and field experiments are conveniently displayed as apparent resistivity

$$\rho_a(\omega) = \mu_0 \omega |C(\omega)|^2 \tag{3.6}$$

and phase of impedance

$$\phi(\omega) = \arg(Ex / Hy) \tag{3.7}$$

as a function of period or frequency $f = \omega/2\pi$. For a homogenous halfspace of resistivity $\rho = 1/\sigma$, the apparent resistivity $\rho_a = \rho$, and the phase is 45 deg. Weidelt (1972) discovered the Kramers-Kroenig relationship between the real and imaginary part of C, also the functions $\rho_a(T)$ and $\phi(T)$ are not independent of each other.

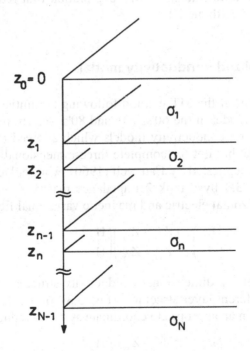

Fig. 3.2 Layered half-space model (Wait 1954)

Schmucker (1973) showed that for periods 6,8,12, 24h or longer, C can be estimated from purely magnetic data of a global induction process, using either the magnetic daily variation or magnetic storms. The MT method is therefore complementary to earlier approaches to the global induction process in a spherical conductivity Earth model (Lahiri and Price 1939). The real part of the complex Schmucker-Weidelt transfer function C is the central depth of the in-phase induced currents and an upper limit of the depth range of the exploration technique. It depends on period and conductivity (Eq. 3.2a). Geomagnetic variations in the 10s - 1000s period range are typically used in studies of conductors in the middle crust, although a wider period range that includes smaller and larger penetration depths is often desirable.

It should be noted at this stage that (Eq. 3.1) is a linear differential equation. Although (Eq. 3.4) is a non-linear relationship between MT models and MT data, even layered earth models which include a very heterogeneous or fractal conductivity structure in one layer will create very smooth ρ_a (T) and ϕ (T) profiles. This is a consequence of Weidelt's (1972) relationships which predicts neighbored MT data for neighbored evaluation periods. The route to fractals in MT exploration first requires a departure from the layered earth model.

3.4 Generalized conductivity model

The development of the MT method following the initial one-dimensional (layered earth) model in the 60's, 70's and 80's was strongly influenced by a series of general conductivity models which allowed for more structure than a 1D model but not for complete three-dimensionality. The first general model was suggested by Cantwell (1960), who replaced the scalar impedance in Eq. 3.3 by a rank 2 impedance matrix Z which links the 2-component horizontal electric and magnetic variational fields E, H

$$\begin{pmatrix} E_x \\ E_y \end{pmatrix} = \begin{pmatrix} Z_{xx} & Z_{xy} \\ Z_{yx} & Z_{yy} \end{pmatrix} \begin{pmatrix} H_x \\ H_y \end{pmatrix} \qquad (3.8)$$

In the case of a 2-dimensional conductivity structure, e.g. two quarter-spaces with different layer structures (Fig. 3.3), or a dyke, this impedance matrix would - in an appropriate coordinate system - reduces to

$$\begin{pmatrix} E_x \\ E_y \end{pmatrix} = \begin{pmatrix} 0 & Z_{xy} \\ Z_{yx} & 0 \end{pmatrix} \begin{pmatrix} H_x \\ H_y \end{pmatrix} \qquad (3.9)$$

with Z_{xy} and Z_{yx} being the different impedances of two decoupled systems of equations describing induction with electric fields parallel and perpendicular to the "strike" of the 2D structure (Swift 1967). The large success of this model was partly due to the fact that numerical solutions for induction in 2D structures were available very early (e.g. Jones and Pascoe 1971). Swift (1967) also provided a scheme for estimating the strike direction of the 2D structure from the elements of the measured impedance matrix, as well as a misfit parameter "skew" which indicates whether the general model explains measured data (see also Vozoff 1972).

In an alternative approach, Larsen (1975) considered the superposition of a regional 1D layered Earth model and a small scale structure of anomalous conductance at the surface (Fig. 3.4). If the size of that anomaly is small compared to the penetration depth p of the electromagnetic field, then impedance matrix of Larsen's general model can be described as

$$\begin{pmatrix} E_x \\ E_y \end{pmatrix} = A \begin{pmatrix} 0 & Z \\ -Z & 0 \end{pmatrix} \begin{pmatrix} H_x \\ H_y \end{pmatrix} \tag{3.10}$$

where Z is the regional impedance of Cagniard's 1D model and A is a real distortion matrix which describes the galvanic (rather than inductive) action of the local scatterer on the electric field. Numerous suggestions have been made for estimating the elements of this distortion matrix from independent information (see review by Groom and Bahr (1992) and references therein).

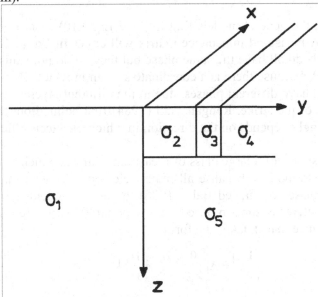

Fig. 3.3 General two-dimensional model (Swift 1967). The conductivity varies only in the y and z directions, and x is the direction of 'strike'

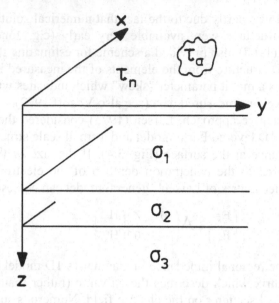

Fig. 3.4 Layered half – space   superposition model (Larsen 1977). A small-scatterer with anomalous conductance τ_a causes amplitude and direction changes of the induced electric fields (after Simpson and Bahr 2005)

At first, the general models Eq.(3.9) and Eq.(3.10) seem to be complementary. A measured impedance matrix will either fit Eq. (3.10), because its elements do all have the same phase but they all do not vanish or it will fit Eq. (3.9) because there is a coordinate system in which, $Z_{xx}= Z_{yy}=0$, but Z_{xy} and Z_{yx} have different phases, due to the different layered structures on both sides of the strike. Ranganayaki (1984) first pointed out that the MT phase strongly depends on the direction in which the electric field is measured.

The existence of a large class of measured matrices which would fit neither of these models, because all matrix elements do not vanish and two different phases occur, led Bahr (1988) to another general model. In it a surface scatterer is superimposed on a regional 2D structure (Fig.3.5) and the impedance matrix takes the form

$$\begin{pmatrix} E_x \\ E_y \end{pmatrix} = A \begin{pmatrix} 0 & Z_{xy} \\ Z_{yx} & 0 \end{pmatrix} \begin{pmatrix} H_x \\ H_y \end{pmatrix} \tag{3.11}$$

This general model was adopted by Groom and Bailey (1989), Bahr (1991), Chave and Smith (1994), and Pracser and Szarka (2000) with improved mathematical techniques for the estimation of the regional strike, and applied to many data sets. Because all strike estimation techniques require the existence of two different phases in the impedance matrix (Berdi-chevky 1999), it was surprising that this general model was successful in most cases: within a wide period range and in most target areas, there is always a magnetotelluric strike. With respect to fractal conductivity structures, the existence of this strike turned out to be the key observation of magnetotelluric exploration of the crust.

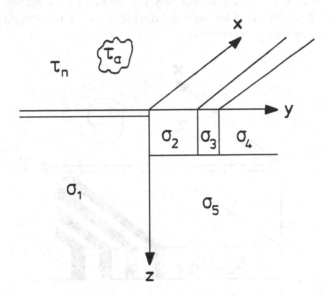

Fig. 3.5 Two-dimensional superposition model (Bahr 1988): Combination of the two dimensional structure in Fig. 3.3 and the anomalous conductance scatterer in Fig. 3.4 (after Simpson and Bahr 2005)

3.5 Conductivity of the middle crust and its anisotropy

The two narrative strings of this chapter, general conductivity models and conduction in the crust, were merged by Kellet et al. (1992) who suggested that the existence of a 'strike' and the phase difference associated with it can be a consequence of electrical anisotropy in a particular depth range. The Eq. (3.11) was still used by Kellet et al. (1992) but their physical model was a superposition of a surface scatterer and an electrically anisotropic regional structure (Fig. 3.6). Although developed for a limited target area, their model can be used as a general one, because it explains why Eq. (3.11) describes so many different data sets. An example for the phase split created by an anisotropic structure is presented in Fig. 3.7. The concept of lower crustal electrical anisotropy in connection with the impedance matrix decomposition was used by Jones et al. (1993) and Eisel and Bahr (1993), and resistivity ratios of up to 60 were used. Bahr et al. (2000) suggested a general 'anisotropy test' based on the magnetotelluric phase in array data, and showed that this test can distinguish between the model of crustal anisotropy and the model of an isolated conductivity anomaly. The search for the origin of the anisotropy was linked to the search for a conduction mechanism, and led to the consideration of two-phase structures with fractal geometry in the crust.

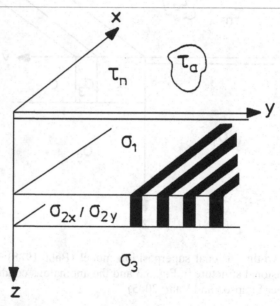

Fig. 3.6 Anisotropic superposition model (Kellet et al. 1992). In numerical studies, the anisotropic layer has either been modeled with two different conductivities σ_{2x}, σ_{2y} (Yin 2000) or with a series of macroscopic conductive lamellae (Tezkan et al. 1992). From magnetotelluric field data, we can not distinguish between these two models (Eisel and Haak 1999), and in the context of this review both are oversimplifications (after Simpson and Bahr 2005)

3.6 Anisotropy and conduction mechanism

The debate on the conduction mechanism in the crust reflected the fact that, besides the rock matrix, an additional conductive component is required to provide the high bulk conductivities observed by magnetotellurics.

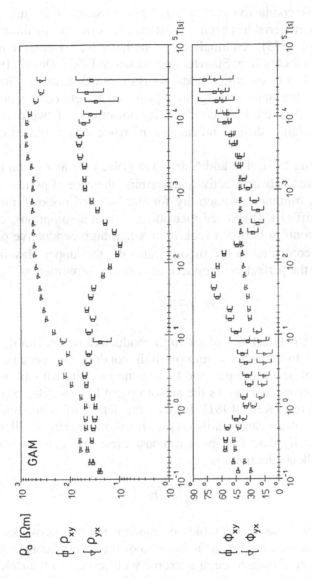

Fig. 3.7 Magnetotelluric apparent resistivity and phase at site GAM in southern Germany (Bahr et al. 2000). The splitting of the two apparent resistivities ρ_{xy}, ρ_{yx} of the two polarizations at periods longer than 10 s is associated with a splitting of the phases ϕ_{xy}, ϕ_{yx}. This splitting occurs at 7 sites of an array in the same period range, suggesting that electrical anisotropy is a more likely explanation than an isolated conductivity anomaly: in the x direction, a conductor is seen which does not occur in the y direction (Bahr et al. 2000)

The bulk conductivity of a very large system of fractures filled with conductive material has been calculated with effective medium theory (e.g. Bruggeman 1935), renormalization methods (e.g. Bernasconi 1978) and percolation theory (e.g. Stauffer and Aharony 1992). David (1993) pointed out that network or critical path analysis can describe a porous medium better than the older mixing laws provided by effective medium theory. With respect to the bulk conductivity anisotropy of two-component systems, percolation theory and random network studies turned out to be the key tools.

According to Hashin and Shtrikman (1962) for any given mixing ratio β, a maximal bulk conductivity describing the case of perfect interconnection and a minimal conductivity for the 'isolated pockets' model are obtained. Waff (1974) showed that under certain assumptions, e.g. the conductivity contrast between rock matrix and high conductive phase should be large compared to the mixing ratio β , the 'upper Hashin-Shtrikman bound' for the perfect interconnection case can be written as

$$\sigma_b = \frac{2}{3}\beta\sigma_m \qquad (3.12)$$

if σ_m is the conductivity of the high conductive phase. But this case is inappropriate to describe anisotropic bulk conductivity, because the mixing ratio cannot be anisotropic, and there is no material with an intrinsic electrical anisotropy as large as the anisotropy in MT models, except for crystalline graphite (Kelly 1981). Instead, the degree of interconnection of the conductive phase can be anisotropic. Interconnectivity smaller than 1 can be numerically described by a dimensionless connectivity coefficient Io, and the bulk conductivity is

$$\sigma_b = \frac{2}{3}\beta\sigma_m Io \qquad (3.13)$$

Bahr (1997) used ensembles of random resistor networks in order to calculate the connectivity as a function of the model geometry. These networks can be chosen to be anisotropic with respect to their electrical connectivity, and the ratio of the connectivities in two different directions increases with increasing complexity of the network. The complexity was increased by the use of embedded networks (Fig. 3.8). Embedded networks were originally suggested by Madden (1976), who nevertheless did not perform random resistor experiments. Fig. 3.9 shows the electrical connectivity Io as function of the percentage p of open fractures. Io was estimated by use of a statistical approach. For every percentage p of open fractures, 1000 embedded network where created. A random generator was used to

distribute low resistance resistors (open fractures) and high resistance resistors (closed fractured) in an actual realization of the embedded network.
The "electrical connectivity" is the average normalized current passing the
network (Bahr 1997). If, for example, p = 0.25 then Io(p) = 0, because
25% open fractures (black in Fig. 3.8) will not form a conducting path
through the entire network. In the vicinity of the percolation threshold, the
Io(p) function has a non-linear increase. Due to the limited size of the network, this numerical approach does not provide the exact solution for the
percolation threshold of the 2-dimensional quadratic network, p = 0.5
(Stauffer and Aharony 1992). The non-linearity of the Io(p) function is the
key to the understanding of the anisotropy: suppose there are slightly more
open fractures in one horizontal direction then in the other, and both p values are in the vicinity of the percolation threshold (Fig. 3.9). The resulting
connectivities in the two directions differ very much. This model can explain how cracks, which are filled with some highly conductive material
and which occur slightly more often in one direction can create anisotropy
of the bulk conductivity.

3.7 The route to fractals: random resistor experiments

The example of a random resistor network in Fig. 3.8 is still too regular to
be a fractal. In the "fractal embedded network" (Fig. 3.10) some sub-
networks are replaced by single resistors, matching a situation where
cracks of many different sizes co-exist in the crust. If many realisations of
this type of network are generated in a statistical approach, then a distribution function of the parameter 'electrical connectivity' is obtained (Fig.
3.11). The variability of the bulk resistivity in the field data in Fig. 3.1
arises from the variability of the connectivity in the highly conductive
component.

No suggestion has been offered by the fractals approach about the origin
and nature of the high conductive material. But the result in Fig. 3.11 can
only be obtained if the system is close to the percolation threshold. If, instead, p is chosen close to 1, then Io (p) will also be close to unity, and
only little direction dependence of Io (p) can be obtained. As a consequence, Io is significantly smaller than 1, and therefore the fraction β of
conductive material has to be enlarged in order to give the same bulk conductivity in Eq. (3.13). This rules out electrolytic conduction in some
cases, because the required porosities become unreasonably high (Bahr
2000, Simpson 2001).

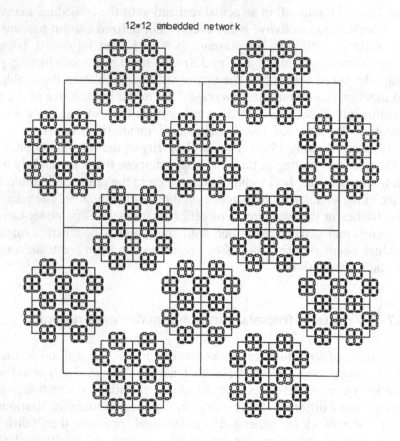

Fig. 3.8 Embedded network with limited connectivity. Black resistors represent high conductive material in open fractures and open resistors represent absence of conductive material. Distribution of 50% 'black' resistors over thew $12^3 = 1728$ possible positions in the network by use of a random generator

The self-similar geometry of the conductive structures in Fig.s 3.8, 3.10 is, of course, a consequence of the employment of embedded networks. But do the anisotropy data and conductivity distribution functions require the self-similarity? Blome (2004) showed that it is not necessary to enforce a self-similar geometry of the model network by employing embedded networks. Blome (2004) performed bond percolation experiments in ordinary (not embedded) 240*240 resistor networks with a systematic variation of the relative number p of 'black' resistor representing the highly conductive phase.

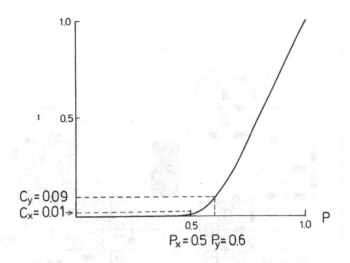

Fig. 3.9 Connectivity Io of the embedded network plotted as a function of the proportion p of conductive material. Suppose that 50 % of the resistors in the x direction and 60% of the resistors in the y direction represent conductive material and therefore the network is slightly above the conduction percolation threshold in both directions. The resulting connectivities Io_x, Io_y in both directions are small compared to the upper Hashin-Shtrikman limit $Io = 1$ and they differ by a much larger factor then p_x and p_y, giving rise to anisotropic conduction in the network

Only at the percolation threshold the model network can exhibit strong anisotropy of the electrical connectivity. Fig. 3.12 shows an example of a resistor network at the percolation threshold. It is well known (e.g. Stauffer and Aharony 1992) that only in the vicinity of the percolation threshold, e.g. p=0.5 for bond percolation in squared lattices, the size-frequency distribution function of clusters of linked bonds is fractal. For smaller p, the large clusters are missing and for larger p only large clusters occur. Thus, from all possible two-component structures, only the ones at the percolation threshold exhibit fractal geometry and explain the statistical properties of the em field data with penetration depth relevant to the lower crust. Recently, Everett and Weiss (2002) showed that electromagnetic responses from near-surface structures can also be fractal signals.

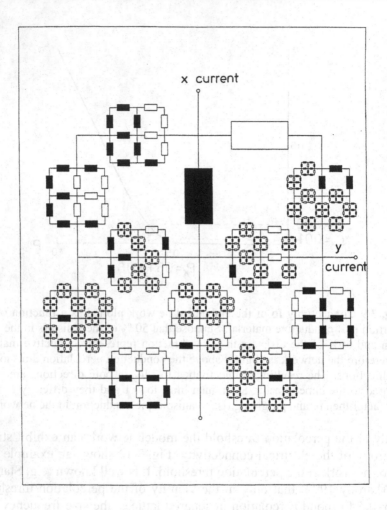

Fig. 3.10 A possible realization of the twice-embedded fractal network. The probability that a resistor is replaced by a small-scale network is 70%

3.8 Discussion: fractals and anisotropy

Kozlovskaya and Hjelt (2000) stress the need for model parameterization schemes that can be used to model the distribution of more than one physical parameter. They choose the fractal rock model because it allows the description of the real complicated rock microstructure by a small number of parameters, and because elastic and electric properties can be calculated within that framework. Here I showed that a statistical evaluation of mag-

netotelluric field data can provide independent evidence that fractal structures exist in parts of the crust. The model which simulates the statistical
properties of the field data is a resistor network in the vicinity of the percolation threshold. In such a network, the resistors representing the conductive material form a fractal cluster system.

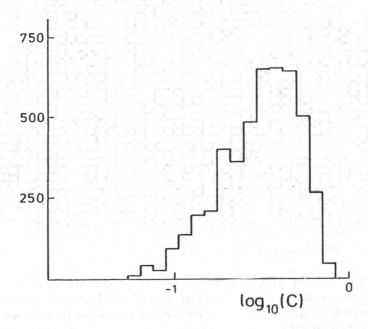

Fig. 3.11 Modeled distribution function of connectivity f (Io) of the twice-
embedded fractal network

   According to Gavrilenko and Gueguen (1989) fluid pressure is the driv-
ing force for hydraulic fracturing. This force is turned off when the hydraulic percolation threshold is reached. The resulting network of fractures
will stay close to the percolation threshold and will have a self-similar geometry, as supported by the self-similarity of cracks in rocks found in experimental studies by Schmittbuhl et al. (1995) and Meheust and
Schmittbuhl (2001). If the fractures are filled with brines, the resulting
conductive network also has a self-similar geometry. In an alternative
model, ancient fluid percolation allowed for the precipitation of graphite
and the conduction mechanism in the network is electronic conduction.
There are more similarities between crack networks and the resistor networks that are employed here.

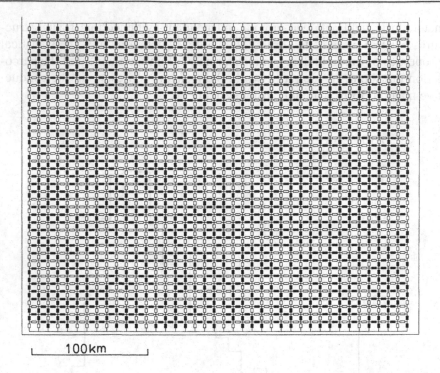

100km

Fig. 3.12 Network of 3200 resistors at the percolation threshold, p=0.5. No fractal geometry is enforced by applying the embedded network approach, but cluster system formed by the resistors representing conductive material has a fractal geometry for p = 0.5

The bulk properties, permeability and electrical conductivity are strongly influenced by the connectivity of cracks (Berkowitz and Adler 1998) and by the connectivity of conductors (Bahr 1997, Kozlovskaya and Hjelt 2000), respectively. Meheust and Schmittbuhl (2001) pointed out that the self-similarity of the crack system creates heterogeneity on all scales, and therefore there exists hydraulic anisotropy on all scales. Electrical anisotropy could occur in crustal conductors on all scales, although the magnetotelluric method can detect it only on the larger scales. In this chapter, I tried to review the route which the magnetotelluric technique went in order to find electrical anisotropy, and to provide some evidence of links between anisotropy and conduction in fractals. Both fields are, however, subject to continuous research, and exciting results should come up in the future.

Acknowledgements

I thank V.P. Dimri for his invitation to write this review and G. Heinson for his hospitality during my Australian summer spent in Adelaide. Also I am thankful to Cambridge University Press for granting their permission to publish figures 5.4 and 5.7.

3.9 References

Allegre CJ, Le Mouel, JL (1993) Introduction to scaling techniques in brittle fracture of rocks. Phys Earth Planet Int 87: 85-93
Bahr K (1988) Interpretation of the magnetotelluric impedance tensor: regional induction and local telluric distortion. J Geophys 62: 119-127
Bahr K (1991) Geological noise in magnetotelluric data: a classification of distortion types. Phys Earth Planet Int 66: 24-38
Bahr K (1997) Electrical anisotropy and conductivity distribution functions of fractal random networks and of the crust: the scale effect of connectivity. Geophys J Int 130: 649-660
Bahr K, Bantin M, Jantos C, Schneider E, Storz W (2000) Electrical anisotropy from electromagnetic array data: implications for the conduction mechanism and for distortion at long periods. Phys Earth Planet Int 119: 237-257
Bailey RJ, Craven JA, Macnae JC, Polzer BD (1989) Imaging of deep fluids in Archean crust. Nature 340: 136-138
Berdichevsky MN (1999) Marginal notes on magnetotellurics. Surveys in Geophysics 20: 341-375
Berkowitz B, Adler PM (1998) Stereological analysis of fracture network structure in geological formations. J Geophys Res 102: 15339-15360
Bernasconi J (1978) Real-space renormalization of bond-disordered conductance lattices. Phys Rev B 18: 2185-2191
Blome M (2004) Heterogenität in der unteren Kruste: Statistische Verzerrungen in magnetotellurischen Array-Messungen modelliert durch perkolative Gleichstrom-Widerstandsnetzwerke. Unpublished Diploma Thesis, Geophysical Institute, University of Göttingen
Bruggeman DAG (1935) Berechnung verschiedener physikalischer Konstanten von heterogenen Substanzen. Annalen der Physik. 5. Folge, Band 24: 636-679
Cagniard L (1953) Basic theory of the magnetotelluric method of geophysical prospecting. Geophysics 18: 605-645
Cantwell T (1960) Detection and analysis of low frequency magnetotelluric signals. Ph.D. thesis. Dept. Geol. Geophys. M.I.T., Cambridge, Mass
Chave AD, Smith JT (1994) On electric and magnetic galvanic distortion tensor decompositions. J Geophys Res 99: 4669-4682
Cull JP (1985) Magnetotelluric soundings over a Precambrian contact in Australia. Geophys J R A Soc. 80: 661-675
David Ch (1993) Geometry of flow for fluid transport in rocks. J Geophys Res 98: 12267-12278
Doyen Ph M (1988) Permeability, conductivity, and pore geometry of sandstone. J Geophys Res 93: 7729-7740
Eisel M, Bahr K (1993) Electrical anisotropy under British Columbia: Interpretation after magnetotelluric tensor decomposition. J Geomag Geoelectr 45: 1115-1126

Eisel M, Haak V (1999) Macro-anisotropy of the electrical conductivity of the crust: a magneto-telluric study from the German Continental Deep Drilling site (KTB). Geophys J Int 136: 109-122

ELEKTB group (1997) KTB and the electrical conductivity of the crust. J Geophys Res 102: 18289-18305

Everett M, Weiss CJ (2002) Geological noise in near-surface electromagnetic induction data. Geophys Res Lett 29(1): 1010 doi: 10.1029/2001GL014049

Frost BR, Fyfe WS, Tazaki K, Chan T (1989) Grain-boundary graphite in rocks and implications for high electrical conductivity in the lower crust. Nature 340: 134-136

Gavrilenko P, Guéguen Y (1989) Percolation in the crust. Terra Nova 1: 63-68.

Gough DI (1986) Seismic reflectors, conductivity, water and stress in the continental crust. Nature 323: 143-144

Greenberg RJ, Brace WF (1969) Archie's law for rocks modeled by simple networks. J Geophys Res 74: 2099-2102

Groom RW, Bailey RC (1989) Decomposition of the magnetotelluric impedance tensor in the presence of local three-dimensional galvanic distortion. J Geophys Res 94: 1913-1925

Groom RW Bahr K (1992) Corrections for near surface effects: decomposition of the magnetotelluric impedance tensor and scaling corrections for regional resistivities: a tutorial. Surveys in Geophysics 13: 341-379

Guéguen Y, Chr David, P Gavrilenko (1991) Percolation networks and fluid transport in the crust. Geophys Res Lett 18: 931-934

Guéguen Y, Palciauskas V(1994) Introduction to the physics of rocks. Princeton University Press, Princeton

Guéguen Y, Gavrilenko P, Le Ravalec M (1996) Scales of rock permeability. Surveys in Geophysics 17: 245-263

Haak V, Hutton VRS (1986) Electrical resistivity in Continental lower crust: In Dawson, JB, Carswell, DA, Hall J, Wedepohl KH, (eds) The Nature of the lower crust, No 24. Geol Soc Spec Publ Blackwell, Oxford, pp. 35-49

Hashin Z, Shtrikman S (1962) A variational approach to the theory of the effective magnetic permeability of multiphase materials. J Appl Phys 33:3125-3131

Heimpel M (1997) Critical behaviour and the evolution of fault strength during earthquake cycles. Nature 388: 865-868

Heimpel M, Olson MP (1996) A seismodynamical model of lithosphere deformation: development of continental and oceanic rift networks. J Geophys Res 101: 16155-16176

Hyndman RD, Shearer PM (1989) Water in the lower crust: modelling magnetotelluric and seismic reflection results. Geophys. JRA Soc 98: 343-365

Jödicke H (1992) Water and graphite in the earth's crust - an approach to interpretation of conductivity models. Surveys in Geophysics 13: 381-407

Jones AG (1992) Electrical conductivity of the continental lower crust: In Fountain, DM, Arculus RJ, RW Kay(eds)Continental Lower Crust, Elsevier, Amsterdam, pp. 81-143

Jones AG, Groom, RW, Kurtz RD (1993) Decomposition and modelling of the BC87 data set. J Geomag Geoelectr 45: 1127-1150

Jones FW, Pascoe LJ (1971) A general computer program to determine the pertur-
    bation of alternating electric currents in a two-dimensional model of a region
    of uniform conductivity with an embedded inhomogeneity. Geophys J R A
    Soc 24: 3-30
Kellet RL, Mareschal M, Kurtz RD (1992) A model of lower crustal electrical ani-
    sotropy for the Pontiac Subprovince of the Canadian Shield. Geophys J Int
    111: 141-150
Kelly BT (1981) Physics of Graphite. Applied Science Publisher, London Engle-
    wood New Jersey
Kirkpatrick S (1973) Percolation and conduction. Rev Mod Phys 45: 574-588
Kontny A, Friedrich G, Behr HJ, de Wall H, Horn EE, Möller P, Zulauf G (1997)
    Formation of ore minerals in metamorphic rocks of the German continental
    deep drilling site (KTB). J Geophys Res 102: 18323-18336
Kozlovskaya E, Hjelt SE (2000) Modeling of elastic and electrical properties of
    solid-liquid rock system with fractal microstructure. Phys Chem Earth A 25:
    195-200
Lahiri BN, Price AT (1939) Electromagnetic induction in non-uniform conduc-
    tors, and the determination of the conductivity of the earth from terrestrial
    magnetic variations. Phil Trans R Soc London (A) 237: 509-540
Larsen JC (1975) Low frequency (0.1-6.0 cpd) electromagnetic study of deep
    mantle electrical conductivity beneath the Hawaiian islands. Geophys JRA
    Soc 43: 17-46
Lastovickova M (1991) A review of laboratory measurements of the electrical
    conductivity of rocks and minerals. Phys Earth Planet Int 66: 1-11
Madden TR (1976) Random networks and mixing laws. Geophysics 41: 1104-
    1125
Mareschal M (1990) Electrical conductivity: the story of an elusive parameter, and
    how it possibly relates to the Kapuskasing uplift (Lithoprobe Canada).In:
    Salisbury, MH, Fountain, DM(eds) Exposed Cross-sections of the Continen-
    tal Crust, Kluwer, Dordrecht, pp 453-468
Mareschal M, Fyfe WS, Percival J, Chan T (1992) Grain-boundary graphite in
    Kapuskasing gneisses and implication for lower crustal conductivity. Nature
    357: 674-676
Marquis G, Hyndman RD (1992) Geophysical support for aqueous fluids in the
    deep crust: seismic and electrical relationships. Geophys J Int 110: 91-105
Meheust Y, Schmittbuhl J (2001) Geometrical heterogeneities and permeability
    anisotropy of rough fractures. J Geophys Res 106: 2089-2102
Pracser E, Szarka L (1999) A correction to Bahr's "phase deviation" method for
    tensor decomposition. Earth Planets Space 51: 1019-1022
Ranganayaki RP (1984) An interpretive analysis of magnetotelluric data. Geo-
    physics 49: 1730-1748
Rasmussen TM (1988) Magnetotellurics in southwestern Sweden: evidence for an
    electrical anisotropic lower crust? J Geophys Res 93: 7897-7907
Rauen A, Lastovickova M (1995) Investigation of electrical anisotropy in the deep
    borehole KTB. Surveys in Geophysics 16: 37-46

Schmittbuhl J, Schmitt F, Scholz CH (1995) Scaling invariance of crack surfaces. J Geophys Res 100: 5953-5973

Schmucker U (1973) Regional induction studies: a review of methods and results. Phys Earth Planet Int 7: 365-378

Shankland TJ, Ander ME (1983) Electrical conductivity, temperatures, and fluids in the lower crust. J Geophys Res 88: 9475-9484

Simpson F (1999) Stress and seismicity in the lower continental crust: a challenge to simple ductility and implications for electrical conductivity mechanisms. Surveys in Geophysics 20: 201-227

Simpson F (2001) Fluid trapping at the brittle-ductile transition re-examined. Geofluids 1: 123-136

Simpson F, Bahr K (2005) Practical Magnetotellurics. Cambridge University Press, Cambridge UK.

Stauffer D, Aharony A (1992) Introduction to percolation theory. Basingstoke Hauts, UK

Swift C.M (1967). A magnetotelluric investigation of an electrical conductivity anomaly in the southwestern United States. Ph.D. thesis. MIT, Cambridge, Mass

Tezkan B, Cerv V, Pek J (1992) Resolution of anisotropic and shielded highly conductive layers using 2-D electromagnetic modelling in the Rhine graben and Black Forest. Phys Earth Planet Inter 74: 159-172

Tikhonov AN (1950) The determination of the electrical properties of deep layers of the earth's crust. Dokl. Acad. Nauk. SSR 73: 295-297 (in Russian)

Vasseur G, Weidelt P (1977) Bimodal electromagnetic induction in non-uniform thin sheets with an application to the northern Pyrenean induction anomaly. Geophys J R A Soc 51: 669-690

Touret J (1986) Fluid inclusions in rocks from the lower crust. In: Dawson JB, Carswell DA, Hall J, Wedepohl KH (eds) The Nature of the lower crust, Geol Soc Spec Publ No 24 Blackwell, Oxford, pp. 161-172

Vozoff K (1972) The magnetotelluric method in the exploration of sedimentary basins. Geophysics 37: 98-141

Waff HS (1974) Theoretical considerations on electrical conductivity in a partially molten mantle and implications for geothermometry. J Geophys Res 79: 4003-4010

Wait JR (1954) On the relation between telluric currents and the earth's magnetic field. Geophysics 19: 281-289

Wannamaker PE (2000) Comment on The petrological case for a dry lower crust by Bruce WD Yardley, John W Valley. J Geophys Res 105: 6057-6064

Weidelt P (1972) The inverse problem of geomagnetic induction. Zeitschrift für Geophysik 38: 257-289

Wong TF, Fredrich JT, Gwanmesia GD (1989) Crack aperture statistics and pore space fractal geometry of Westerly granite and Rutland quartzite: implications for an elastic contact model of rock compressibility. J Geophys Res 94: 10267-10278

Xu Y, Poe, BT, Shankland TJ, Rubie, DC (1998) Electrical conductivity of oli-
    vine, wadsleyite, and ringwoodite under upper-mantle conditions. Science
    280: 1415-1418
Yardley BWD, Valley JW (1997) The petrological case for a dry lower crust. J
    Geophys Res: 102, 12173-12185
Yardley BWD, Valley JW (2000) Reply. J Geophys Res 105: 6065-6068
Yin C(2000) Geoelectrical inversion for a one-dimensional anisotropic model and
    inherent non-uniqueness. Geophys J Int 140:11-23
Zallen R (1983) The Physics of Amorphous Solids. John Wiley, New York

Chapter 4. Regularity Analysis Applied to Well Log Data

Maurizio Fedi, Donato Fiore, Mauro La Manna

Dipartimento di Scienze della Terra, Università di Napoli , Italy

4.1 Summary

Well logs are largely used for oil exploration and production in order to obtain geological information of rocks. Several parameters of the rocks can be scanned and interpreted in term of lithology and of the quantity and kind of fluids within the pores. Generally the drilled rocks are mostly sedimentary and the modelling is mainly petrophysical. Here we analyze four logs from the KTB Main Borehole, drilled for the German Continental Deep Drilling Program. The hole cuts across crystalline rocks like amphibolites, amphibolite-metagabbros, gneiss, variegated units and granites.

A multifractal model is assumed for the logs and they are analyzed by a new methodology called Regularity Analysis (RA), which maps the measured logs to profiles of Holder exponents or regularity. The regularity generalizes the degree of differentiability of a function from integer to real numbers and it is useful to describe algebraic singularities related not only to the classical model of jump discontinuity, but to any other kind of 'edge' variations. We aim at a) characterizing the lithological changes of the drilled rocks; and b) identifying the zones of macro and micro fractures. The RA was applied to several geophysical well logs (density, magnetic susceptibility, self potential and electrical resistivity) and allowed consistent information about the KTB well formations. All the regularity profiles independently obtained for the logs provide a clear correlation with lithology and from each log we derived a similar segmentation in terms of lithological units.

A slightly different definition of regularity, called an average-local regularity, yields a good correlation between each known major fault and local maxima of the regularity curves. The regularity profiles were also compared with the KTB "Fracture Index" (FI), showing a meaningful relation among maxima of regularity and maxima of fracture index.

4.2 Introduction

Well log prospecting is mainly used for characterizing reservoirs in sedimentary rocks; in fact it is one of the most important tools for hydrocarbon research among international oil companies. Less frequently, but with an increasing interest, other kinds of rocks (i.e. metamorphic and magmatic rocks) have been drilled, like in the German Deep Drilling Project (KTB) and in the International Ocean Drilling Project (ODP). This kind of deep drillings may help to fix some features about the structural and geological setting of an area. Crystalline rocks have a complex mineralogy and a low porosity and permeability, typically decreasing with the depth. The occurrence of macrofractures and related systems of microfractures (Gregg and Singh 1979) may however account for an increase of porosity at depths allowing fluid transport over large distances.

Following Bremer et al. (1992) we assume here that log signals will depend on lithology (rock matrix) as well as on macro and microfractures. It is often assumed that geological units are marked by sharp boundaries, so that most of the research for interpreting logs has regarded the possibility of detecting the edges of such boundaries (Vermeer and Alkemade 1992). The edges are typically regarded as jump (or step) discontinuities, which mark discontinuities of piece-wise continuous functions which are locally homogeneous.

Here, we assume a more general definition for edges, which extends over the complexity of piece-wise functions and considers general algebraic singularities instead of jump discontinuities only. In the case of an isolated singularity we will refer to a local scale-invariance. A multiscale local analysis is very useful to study the isolated singularities since these may be characterized by a parameter, called Holder regularity. It may be meaningful thought as a generalization of the degree of differentiability from integer to real numbers (see section 4.5). If the singularities are dense, or accumulated, the medium has to be considered as a fractal and the local multiscale analysis becomes much more complicated, due to the reciprocal interferences from the various singularities.

Homogeneous fractals, or simply monofractals, are however characterized by a complexity which is well accounted for just a single global parameter, the fractal dimension, which is closely related to a single degree of Holder regularity. An efficient estimator of the fractal dimension is the Fourier or wavelet power spectrum. Multifractals, instead, have a variable degree of Holder regularity. This means that the scale-dependence is inhomogeneous, so that a distance dependent frequency (or scale) analysis is needed. The optimal tool is given by wavelets, being well localized both in distance and frequency. Self-similarity and the concept of multifractals in geophysics have been pioneered by Mandelbrot (1982). Since then, fractal models have arisen often in many scientific disciplines, such as physics, chemistry, astronomy and biology.

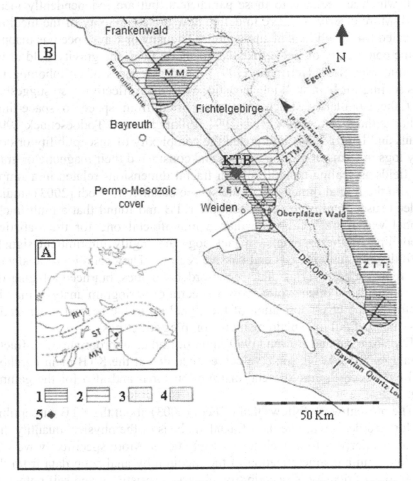

Fig. 4.1 A. Scheme of main units of Hercynian basement (Kossmat 1927 modified). RH-Rhenohercinian Zone; ST-Saxothuringian Zone; MN- Moldanubion Zone. B. Main tectonometamorphic units and DEKROP seismic line. 1 metamorphic nappes; 2. lower nappes; 3.ST; 4. MN; 5. KTB well. MN-Munchberg Massif; ZEV-Erberdorf-Vohenstrauss Zone; ZTT-Tepl-Taus Zone; ZTM- Tirshenreuth-Maharing Zone

The self-similar fractals show an irregular structure at any scale, which is similar at any zooming (in and out) of the signal. Real phenomena may be rarely described using simple deterministic fractal models, but similarity can hold on several scales in a statistical sense, leading to the notion of random fractals. A random fractal model is assumed here for well log data.

An important issue is linking in a coherent way the complex behaviour of the physical properties measured with well logs to other physical quantities, which are related to those parameters, but are independently measured. More clearly we have to define how the complexity of the medium, revealed from well logs of susceptibility, density or wave speed, is mapped to the complexity of fields, like the magnetic field, the gravity field or the seismic wavefield. Herrmann (1997) found evidences of an inhomogeneous scaling for both well log acoustic waves and reflectivity, so suggesting that the singularity structure is transported from space to space-time. Other authors (Gregotski et al. 1991, Pilkington and Todoeschuck 1993, Maus and Dimri 1994) interpreted the complexity of susceptibility or density logs in terms of scaling sources and considered their magnetic or gravity fields as scaling quantities, with fractal dimensions related in a simple way to the fractal dimension of the source parameters. Fedi (2003) studied a deep susceptibility log data from the KTB and found that a multifractal model was more appropriate than a monofractal one for the statistical modelling of these signals. In fact, logs are highly intermittent signals, with distinct active bursts and passive regions. They cannot be satisfactorily represented in terms of a second-order statistics, but need a higher order statistics. In other words, power spectra or variogram analysis may be useful for the characterization of the signal up to the second order statistics, but may fail in explaining more complex structures.

Even Marsan and Bewan (1999) did not find an appropriate monofractal model for the P-wave sonic velocities recorded at the KTB main borehole and evidenced the multifractal distribution for it and also for the gamma log.

The present work follows that of Fedi (2003) about the KTB susceptibility log and deals with the multifractal analysis of the physical quantity distributions derived from well log measurements. More specifically we will try to obtain a characterization of the medium by analyzing data from the KTB logs of density, susceptibility, electrical resistivity and self potential. The multiscale structure derived from logs will be compared with each parameter and also to the known distribution of macro and microfractures and lithology, following a technique (Fedi et al. 2003) for time varying geomagnetic signals. We aim to improve the classical rock characterization by performing a multifractal analysis for more than a single log.

4.3 KTB: an example

In agreement with Vollbrecht et al. (1989) the Central Europe may be divided into four main tectonic units: the Rhenohercynian zone (RH), the Saxothuringian zone (ST), the Moldanubian zone (MN) and Subvariscan Foredeep. In this region many units are also present such as klippen or metamorphic nappes, forming the uppermost structural level (Behr et al. 1984), which are the Munchberg Massif (MM), the Erberdorf-Vohenstrauss Zone (ZEV) and the Tepl-Taus Zone (ZTT) (Fig. 4.1).

The KTB well logs studied in this work fall under the ZEV unit. A geological sequence of the KTB main borehole area is shown in Fig. (4.2).

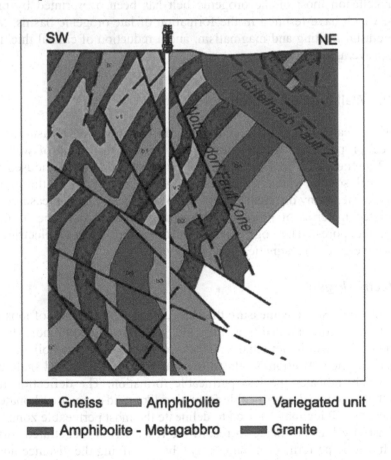

Gneiss   **Amphibolite**   **Variegated unit**

Amphibolite - Metagabbro   **Granite**

Fig. 4.2 Stratigraphic sequence of ZEV units from KTB well (from http://icdp.gfz-potsdam.de/html/ktb, modified)

A broad, sinuous, roughly E-W trending Variscan fold belt represents one of the most important geological features of the Europe. The Variscan Orogen of Northern Europe consists in a series of Ordovician to Carboniferous rift-generated basins, separated by metamorphosed crystalline ridges, which were progressively closed by the northward migration and subsequent collision of African Gondwanaland with northern Baltica. This process has given rise to a system of major and minor fold-and-thrust belts on the external flanks of the entire system as well as to a major zone of highly metamorphic rocks in the internal core. The European Variscides are a prime example of a fossil, deeply eroded collisional mountain belt, which has involved sequential collision of several continental plate fragments. After collision most of the orogenic belt has been overprinted by processes, which have resulted in the formation of late-orogenic basins, large scale crustal melting and magmatism, and a reduction of crustal thickness back to average thickness.

4.3.1 Well logs measurements

Boreholes can yield much geological information by measuring the physical properties of the penetrated formation with the aid of wire line logs. The equipment that measures the physical properties is housed in a cylindrical sonde, which is lowered down to the borehole through an electric cable. Many different parameters of the rocks can be measured and interpreted in terms of lithological porosity, quantity and type of fluids within the pores. The logs studied in this work are (1) electric, (2) radioactive, and (3) magnetic susceptibility.

Electric logs

Self-Potential (SP) logs measure the charge caused by the flow of ions Na^+ and Cl^- from concentrated (generally the formation fluids) to more diluted solution. The electric potential is strictly related to the permeability of the formation. The deflection is related to an arbitrarily determined shale base line, which indicates the less permeable formation. The deflection indicates the presence of a permeable formation like sandstone or carbonate. In practice, the SP log may be used to delineate the most permeable zone.

Resistivity logs are made to record the resistivity of penetrated formation. It can be performed in many ways, by modifying the distance and/or the configuration of the electrodes on the sonde. So we can have conventional resistivity logs, laterologs, which focus the current horizontally, and induction logs that use a variable magnetic field to generate current in the formation.

Radioactivity logs

Gamma ray logs measure the natural radioactivity of a formation. The main radioactive element of sedimentary rocks is potassium, generally present in illitic clays, feldspars, mica and glauconite. The radioactive minerals existing in the organic matter (abundant in oil shale) are derived from the Uranium and Thorium series. Radioactivity is measured in API units. Recent gamma ray spectrometry allowed differentiation among various radioactive minerals.

Neutron logs are made by a neutron bombardment of the penetrated formation that emits gamma rays in proportion to their hydrogen content. The response of neutron log is essentially correlated to the porosity.

Density logs and gamma-gamma logs measure the formation density by emitting gamma rays and recording the gamma rays radiating from the formation; the recorded gamma rays can be related to the electron density of the atoms in the formation, which is related to the bulk density of the formation.

Magnetic susceptibility log

This log can be used in rocks presenting a noticeable magnetic susceptibility to infer lithological changes or alteration zones.

Crystalline rocks are generally considered very tight with low porosity and the fluid transport may be possible only in fractured zones. In particular, for the KTB crystalline rocks macrofractures are normally surrounded by a zone of microfractures (Bremer et al. 1992) and more than of 50% of the pores can be attributed to this kind of fractures, also called as capillaries. So, the response of logs in crystalline rocks is not only in term of rock matrix, permeability and porosity etc., like in sedimentary rocks, but also it depends on the degree of fracturization.

4.4 The regularity analysis

As shown in several studies it is quite reasonable to assume a multifractal model for well log signals (Herrmann 1997, Marsan and Bean 1999, Fedi 2003). We assume that the log signals will depend on lithology (rock matrix) as well as on the occurrence of macro and microfractures (Bremer et al. 1992). Our aim is therefore at assessing whether the properties of the lithology and of the fractures may be determined not only from the measured well logs, but even from the Holder regularity of such logs. The analysis of the Holder regularity will be hereafter called Regularity Analysis (RA) similarly to the case studied for time geomagnetic anomalies (Fedi et al. 2003).

RA detects the second order statistical properties of the analysed signals, helping to distinguish, for instance, among signals characterized by the same power spectrum. An example of this is represented by two random processes: the fractional Brownian motion (fBm) and the general Fractal noise (gFn). Even though such processes show comparable power spectra, they are strongly different. The fBm is a typical monofractal process, while the fGn shows clearly multifractal characteristics, like inhomogeneous spikiness or, in other words, a various degree of intermittence with sudden bursts of high frequency activity and large outliers. Multifractals were introduced by Mandelbrot (1974) to describe the turbulence phenomena; subsequently they have been used in many different contexts. The typical construction of a multifractal process or measure can be obtained iteratively in a multiplicative way from a coarse scale and developing the details of the process on finer scales. An example of this is the binomial multifractal (Feder 1988). There are two ways to study the properties of multifractals, which are respectively of local and global nature (Fedi 2003). The first one consists of defining a procedure to estimate the global repartition of the various Holder exponents, but not their location. As such, it leads to the definition of a spectrum, $D(\alpha)$, called singularity spectrum, which is useful to assess the multifractal properties of a given signal.

The second one is related to the possibility of estimating the Lipschitz-Holder regularity locally. This is possible especially if the singularities appear isolated. Otherwise, the estimation may be difficult, due to interference effects, which may occur especially at large scales and also to the finite numerical resolution. In both cases suitable techniques are based on the continuous wavelet transform (Mallat 1998, Flandrin 1999) and, in particular, on the wavelet transform at local maxima, as described in a landmark paper of Mallat and Hwang (1992). The regularity analysis is based on this second approach. In the well log case, one may indeed search for depth regions, defined by different lithologies, scaling in its own way, and try to characterize the intricate set of not isolated singularities, provided their scaling behaviour persists over a wide enough scale range.

Strictly speaking, the singularities (i.e. (ir) regularities) of a signals $f(x)$ are the points at which the derivative of a given function of a complex variable does not exist but every neighbourhood of which contains points for which the derivative exists.

A function f(x) has a local Holder exponent α at x_0 if and only if with constants C, $h_0 > 0$ and a polynomial P_n of order n such that for $h < h_0$:

$$\left| f(x_0 + h) \; P_n (h) \right| \leq C \left| h \right|^{\alpha} \qquad\qquad n < \alpha < n+1 \qquad (4.1)$$

The Holder regularity of a given signal at some point x_0 is the superior bound of all α verifying the above equation. If the Holder regularity is $\alpha < 1$ at some point x_0, the signal is not differentiable at x_0 and α will fully characterize the singularity type. However, for tempered distributions, also negative exponents may be considered (Mallat and Hwang 1992), similar to a Dirac $\delta(x)$ distribution.

In this chapter, we will consider two different approaches for computing the regularity of a multifractal signal. As mentioned earlier the first one is the local approach, based on the estimation of regularity at each point, and the second one is the average-local approach, which is based on the estimation of the regularity within a moving window. In both cases, the first step is the computation of the Continuous Wavelet Transform (CWT) of the signal. The Continuous Wavelet Transform of $f(x) \in L^2 (R)$ at scale s and point x_0 is

$$Wf(x_0, s) = \left\langle f, \psi_{x_0, s} \right\rangle = \int_{-\infty}^{+\infty} f(x) \cdot \frac{1}{\sqrt{s}} \cdot \psi \left(\frac{x - x_0}{s} \right) dt \qquad (4.2)$$

where the kernel function $\psi_{x_0 s}(x) = 1/(s)^{-1/2} \psi[(x - x_0)/s]$ is the analyzing wavelet.

The choice of the analyzing wavelet is significant in order to obtain a good computation of the CWT. First of all, the analysing wavelet must possess a number of vanishing moments adequate to analyze the given signal and then it has also to be more regular (or smoother) than the process under study. Otherwise, the analysis will be biased by the wavelet's properties instead of the signal.

The estimation of the regularity α can be obtained by the CWT considering the local modulus maxima of the CWT at the singularity point x_0. The regularity α is evaluated by the relationship $\gamma = 2\alpha + 1$, where γ is the log-log slope of the amplitude of the modulus maxima line with respect to the scale.

The average-local approach, instead, provides the regularity α using a moving window L. In practice, it yields the estimation of the regularity (Holder exponent α) at each point x_i of the signal computing the global regularity in the ith moving window centered in x_i. Such global regularity is evaluated again by the relationship $\gamma = 2\alpha + 1$ where γ is now the log-log slope of the component at scale s of the CWT power spectrum related to the ith window, with respect to s. The α is referred to the ith window center and represents a smoothed and more stable evaluation of the local regularity.Basically, using these two approaches we have two different levels of study of the problem. In the local one we will study the signals at the finest detail, while using the average-local we are smoothing out much of the detail and retain the information related closely to the main structural phenomena.

4.5 RA applied to well log data

We now discuss the application of the RA technique to the KTB well logs data. We consider logs of density (D), magnetic susceptibility (MS), self-potential (SP) and electrical resistivity (ER).

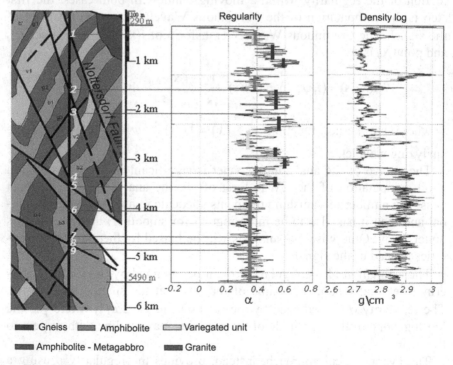

Fig. 4.3 Local regularity of the KTB in which log D is correlated with the geological section. The RA evidences rather homogeneous zones which seem well correlated to the lithologic units marked by the same colours

As described in previous section (4.4), the RA provides an evaluation of the local and average-local regularity of the available well log data. Roughly speaking the local approach provides punctual and somewhat noisy information about the lithological and also other structural effects, like faults, while the average-local approach tends to reveal the occurrence of the major fractured zones. In practice, we will see that the second approach allows a better identification of the fractured zones, reducing the influence of the lithology. In the following, we will discuss the results obtained using either a local or the average-local approach.

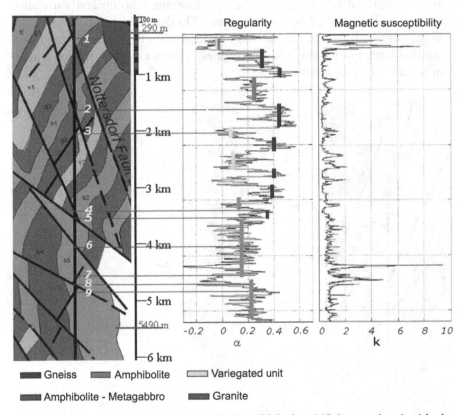

Fig. 4.4 Local regularity of the KTB in which, log MS is correlated with the geological section. Also for MS the RA evidences rather homogeneous zones which seem well correlated to the lithologic units marked by the same colours

4.5.1 Local RA curves

The regularity profiles independently obtained from the logs (Fig. 4.3-4.6), computed using the local approach, present a clear correlation with lithology. While a meaningful segmentation in terms of lithological units was already obtained for the magnetic susceptibility log only (Fedi 2003), here we find further insights by comparing the regularity values for the several analyzed logs. The first zone (from 0 to 3500 m approximately), is relative to a sequence consisting of variable units: gneiss, variegated and amphibolite units. From 3000-6000 m the well intersects a rather homogeneous zone of amphibolite units, sometimes interested by major faults. First of all we note that regularities from density and magnetic susceptibility logs have a very similar behavior. For instance, the highest value of α in both cases is obtained for the gneiss rocks, while amphibolite and variegated units are well identified by lower values. On the other hand, similar considerations occur for the regularities obtained from the resistivity and self-potential logs, but we find that the highest values are obtained for amphibolite and variegated units.

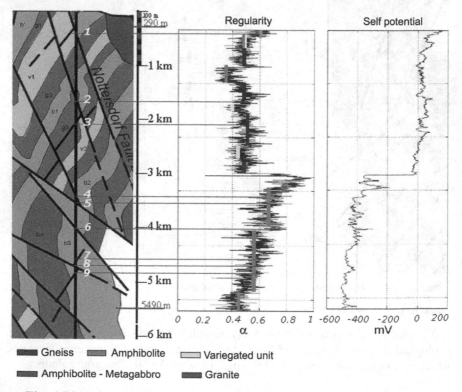

Fig. 4.5 Local regularity of the KTB in which log SP is correlated with the geological section. Also for SP, the RA evidences rather homogeneous zones which seem well correlated to the lithologic units marked by the same colours

The different behavior may be caused by the major role played by porosity in fractured zones for electrical logs with respect to the logs of density and magnetic susceptibility. The values of the Holder exponent that we find consistently range between 0.45 and 0.55 for Gneiss and it is estimated in an interval from 0.3 to 0.45 for the Variegated Units. As already said, regularity of the Amphibolites Units is instead different for the several logs: it is estimated between 0.25 and 0.3 for D and MS logs, while it is between 0.45 and 0.7 for the two electrical logs.

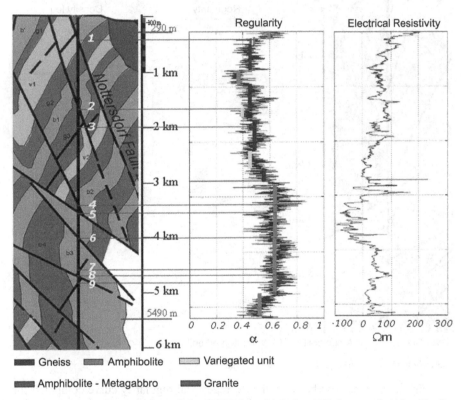

Fig. 4.6 Local regularity of the KTB in which log ER is correlated with the geological section. Also for ER, the RA evidences rather homogeneous zones which seem well correlated to the lithologic units marked by the same colours

4.5.2 Average-local RA curves

Now, we describe the regularity curves computed using the average-local approach. As mentioned earlier, this method provides an estimation of the regularity of the signal by computing the global regularity within a moving window. In order to obtain a suitable regularity evaluation it is important to choose an optimal window length. After several attempts, we selected a window with a length of 40 points to analyze the logs D, MS, SP and ER. This was the minimum length yielding a rather good linearity in a log-log plot between the local CWT modulus maxima versus scales. In this way a reliable evaluation of the regularity exponent α is obtained, since the scaling exponent may be estimated through a linear regression with a good correlation coefficient.

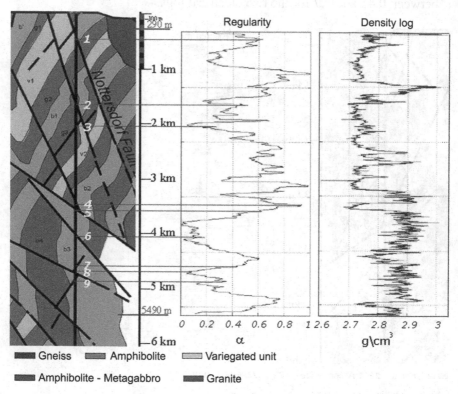

Fig. 4.7 Comparison between the average-local regularity curve computed for log D and the geological section. The red lines connect the structural features (faults) present in the geological section with local maxima of regularity. This shows a good correlation among known macrofractures and local maxima of log D regularity

The average-local regularity curve for four well log data (D,MS,SP,ER)are plotted in Figs 4.7 to 4.10, in which we observe a good relation between each major fault reported in the geological section (numbered from 1 to 9) and local maxima of the regularity curves. This shows that the macrofractures affect regularity in a consistent way, even though the physical effects relating the several measured quantities to the occurrence of macrofractures is not the same. The converse is not strictly true: some regularity maxima do not correspond to any fault. However an interesting comparison could be made using microfracture information. We have also correlated the computed regularities with the KTB fracture index (FI) (Bremer et al. 1992). Their fracture index was intended to be: a) sensitive to micro as well as to macrofractures; b) a better measure of the 'fracturedness' of the rocks than the fracture density of cores.

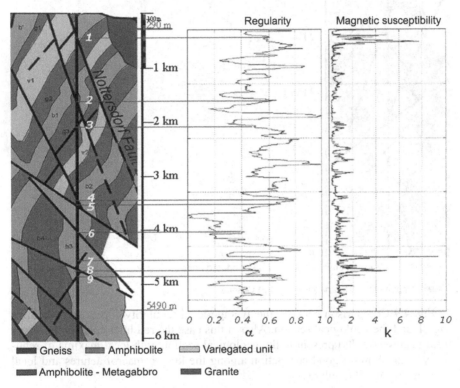

Gneiss   Amphibolite   Variegated unit

Amphibolite - Metagabbro   Granite

Fig. 4.8 Comparison between the average-local regularity curve computed for log MS and the geological section. Also in this case the red lines connect the structural features (faults) present in the geological section with local maxima of regularity. This shows a good correlation among the known macrofractures and local maxima of log MS regularity

A surprising relation between the local maxima of regularity curves and fractured zones identified by the FI is seen in Fig. 4.11. This relation can be qualitatively explained with an increase of the homogeneity of the fractured zone, due to the strong increase of fluids within the enlarged system of pores connected to the fractured zones, which in turn should cause an increase of the regularity. We recall in fact that α=-1 corresponds to spikes and α =1 instead to the bell-shaped singularities (Fedi et al. 2003).

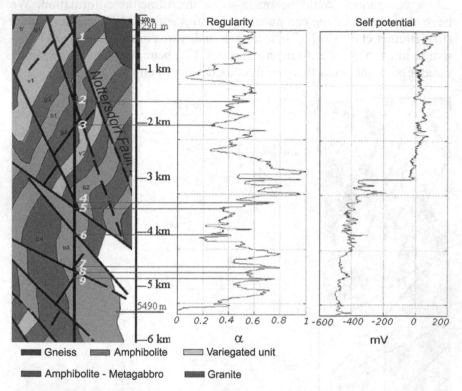

Fig. 4.9 Comparison between the average-local regularity curves computed for log SP and the geological section. Also in this case the red lines connect the structural features (faults) present in the geological section with local maxima of regularity. This shows a good correlation among the known macrofractures and local maxima of log SP regularity

4.6 Discussion

In this chapter, we have proposed a method for analyzing geophysical well logs, based on the application of the regularity analysis to a multiple set of logs. We applied the technique to the data gathered from the KTB well. The well log interpretation is generally based on the interpretation of the several logs measurements, aiming at improving the information related to the geological and structural characterization and to the fluid content of the drilled formations.

Here we focused our attention to the characterization of the signals in terms of its statistical properties at an order higher than two. We assumed a multifractal model for well logs and then applied RA to several logs. Two different approaches were considered for computing the regularity of a multifractal signal. The first is a local approach, based on the estimation of regularity at each point, and the second is an average-local approach, which is based on the estimation of the regularity within a moving window. Our results show that the first approach is more suitable to detect the changes in lithological formations, while the second is more sensible to the occurrence of macro- and microfractures.

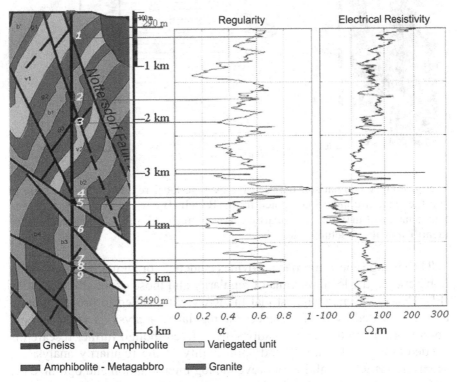

Fig. 4.10 Comparison between the average-local regularity curve computed for log ER and the geological section. Also in this case the red lines connect the structural features (faults) present in the geological section with local maxima of regularity. This shows a good correlation among the known macrofractures and local maxima of log ER regularity

We find a strong correlation between the different regularity profiles. The correlation is more striking between the electrical logs as well as between density and susceptibility logs. So, the analysis of the KTB logs allows us to conclude that the RA provides a rather unique characterization of the drilled rock formations and that just a few well log types may be used to characterize the drilled formations.

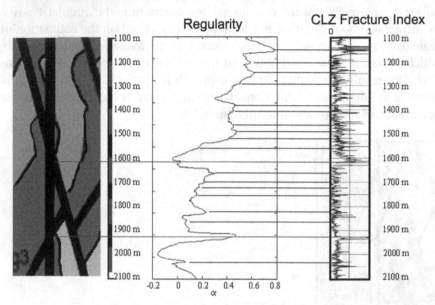

Fig. 4.11 A comparison between the average-local regularity curve of log D and the "Fracture index" computed for KTB well (Bremer et al. 1992). It is seen that many local maxima of regularity curves correspond well to fractured zone, as identified by the Fracture Index

The average-local approach provides evidence for an interesting correlation between the local maxima of regularity and those of the fracture index (FI). Hence, we may suggest that the RA yield detailed information close to that provided by the FI. Hence, the regularity analysis is found to be a good tool for fracture investigations also. Even though further research is needed to assess the utility and applicability of the regularity analysis, it seems nevertheless valid to improve the log interpretation and optimize the time for the analysis and the interpretation of the geological features of stratigraphic sequences.

Acknowledgement

I extend my acknowledgements to Academic Press for their permission to publish a figure on page No. 58 from Elements of Petroleum Geology, 1998, 2nd Ed. by RC Selley.

4.7 References

Behr HJ, Engel W, Franke W, Giese P, Weber K (1984) The Variscan Belt in Central Europe: main structures, geodynamic implications, open questions. Tectonophysics 190:15-40

Bremer MH, Kulenkampff J, Schopper JR (1992) Lithological and fracture response of common logs in crystalline rocks. Geological applications of wireline logs II, Geological Society Special Pubblications 65 pp 221-234

Feder J (1988) Fractals. Plenum Press, New York London

Fedi M (2003) Global and local multiscale analysis of magnetic susceptibility data. Pure Appl Geophys 160:2399–2417

Fedi M, La Manna M, Palmieri F (2003) Non stationary analysis of geomagnetic time sequences from Mt. Etna and North Palm Springs earthquake. J Geophys R 108: n.B10, 2493 10.1029/2001JB000820

Flandrin P (1999) Time frequency/time-scale analysis. Academic Press, London

Gregg JS, Singh KSW (1979) Adsorption surface area and porosity. Academic Press, London

Gregotski ME, Jensen O, Arkani-Hamed J (1991) Fractal stochastic modelling of aeromagnetic data. Geophysics 56:1706-1715

Haykin S (1996) Adaptive filter theory. In: Kalaith T (ed) Info and Sys Sci Ser, Prentice Hall, pp. 365-572

Hermann FJ (1997) A scaling medium representation, a discussion on well-logs fractals and wave, PhD Thesis. Faculty of Applied Physics, Delft University of Technology, Delft, The Netherlands

Mallat S, Hwang WL (1992) Singularity detection and processing with wavelets. IEEE Trans Inf Theory 38: 617-643

Kossmat F(1927) Gliederung des varistischen Gebirgsbaues. Abhandlungen Sächsischen Geologischen Landesamts 1: 1-39

Mallat S (1998) A wavelet tour of signal processing. Academic Press, London

Mandelbrot BB (1974) intermittent turbulence in self-similar cascades; divergences of high moments and dimension of the carrier. J Fluid Mech 62:331-345

Mandelbrot BB (1982) The fractal geometry of nature. W H Freeman, New York

Marsan D, Bewan CJ (1999) Multiscaling nature of sonic velocities and lithology in the upper crystalline crust: evidence from the main KTB borehole. Geophys Res Lett 26:275-278

Maus S, Dimri VP (1994) Scaling properties of potential fields due to scaling sources. Geophys Res Lett 21:891-894

Pilkington M, Todoeschuck JP (1993) Fractal magnetization of continental crust. Geophys Res Lett 20:627-630

Vermeer PL, Alkemade JAH (1992) Multiscale segmentation of well logs. Mathematical Geology 24:27-43

82   Maurizio Fedi, Donato Fiore, Mauro La Manna

Vollbrecht A, Weber K, Schmoll J (1989) Structural model for the Saxothurin-
gian-Moldanubian suture in the Variscan basement of the Oberpfalz (Northeastern Bavaria) interpreted from geophysical data. Tectonophysics 157 :123-
133

Chapter 5 Electrokinetic Effect in Fractal Pore Media as Seismoelectric Phenomena

V.V. Surkov[1], H.Tanaka[2]

1 Moscow State Engineering Physics Institute, Moscow, Russia
2 RIKEN International Frontier Research Group on Earthquakes, Tokai University, Japan

5.1 Summary

In this chapter the theory of electrokinetic effect on fractal is developed. Inhomogeneous water-saturated medium with fractal structure of pore space is a subject of investigation. The fluid migration along the percolation clusters is accompanied by the electrokinetic effect caused by contact potential difference at phase bounds. The electrokinetic current density is found to depend on both the transport critical exponent and correlation length critical exponent. Two different models of the inhomogeneity embedded in rock are considered. In the first model a fractal core with high pore fluid pressure is surrounded by weak permeable rock. In the second one a non-fractal core composed of high permeable broken rock is surrounded by a fractal periphery. The electrokinetic current in the fractal regions results in the appearance of electric currents in conductive layers under the ground. Amplitude of the electric signal versus the size of fractal structure and distance from the source is estimated. This dependence is applied for seismic electric signals (SES) occasionally observed prior to great crustal earthquakes. Interestingly enough the empirical dependence of the SES amplitude on earthquake magnitude can be explained solely by accounting for the scaling arguments. A special credit is paid to study the duration of the SES. The fluid migration is described by a generalized diffusion type equation in Euclidean spaces to fractal spaces. The use of this equation makes it possible to consider the envelope of the fluid pressure, which is, in fact, a non-analytic function. The SES duration is found to be the same order of magnitude as the time of fluid diffusion on fractal that is much greater than the duration of the electric current propagation in conductive medium.

5.2 Introduction

The electrokinetic current in pore water-saturated rock is due to the migration of charged fluid. Walls of pores and cracks can adsorb ions of certain elements from fluid whereas fluid is charged oppositely, and then fluid flow is accompanied by an electric current. Large-scale rock fractures in the vicinity of an earthquake focal zone can be accompanied by the migration of underground fluid, thus the electrokinetic current flows. It has long been known that the electrokinetic effect is responsible for the seismoelectric phenomena associated with seismic wave propagation in moist soil (so-called seismoelectric effect of the second kind) (Ivanov 1940, Frenkel 1944, Martner and Sparks 1959). The similar effect is supposed to be a possible cause of so-called seismic electric signals (SES) observed in seismoactive regions (Varotsos et al. 1984a, b 1996, Uyeda et al. 2000). In this case the large scaled fluid migration is caused by the tectonic stress variations before seismic events. The continuous monitoring of telluric potential difference on the network of grounded non-polarization electrodes have been made to detect the isolate SES with duration from tens minutes up to several hours. It was hypothesized by Varotsos et al. (1984a, b) that such signals occasionally observed several hours or days before an impending strong crust earthquake can serve as a short term electromagnetic precursor. The logarithm of the SES amplitude was found to depend on magnitude of a forthcoming earthquake in a linear fashion. The coefficient proportionality in this empirical dependence does not follow from theoretical models based on electrokinetic effect (Bernard 1992, Fenoglio et al. 1995). Surkov et al. (2002) have shown that some features of this dependence including the coefficient proportionality can be explained on the supposition that the electrokinetic current spreads in water-saturated rock with fractal structure of pore space. The part of the international geophysical community doubts about the validity of the assumption that the SES is related to earthquake. In spite of this fact the electrokinetic effect has been the subject of a great deal of laboratory research in the past two decades (Mizutani and Ishido 1976, Jouniaux and Pozzi 1999). The possibility for the detection of earthquake electromagnetic precursors is an intriguing problem which is widely discussed up to now.

The goals of this chapter are (1) to relate the SES amplitude and duration to critical exponents and fractal zone size and (2) to drive the relations between the SES parameters and magnitude of earthquakes on the basis of the electrokinetic effect on a fractal structure above percolation threshold.

5.3 Estimation of electric field amplitude

In multiphase heterogeneous medium the contact potential difference and electric charges can be formed on the phase bounds. For example, in the pore water-saturated rocks the underground water consists of electrolyte solutions that include ions and dissociated molecules. The surfaces of cracks and pores can adsorb ions of certain sign from the fluid. For example the surface charge of solid phase can be negative due to acid dissociation of the surface hydroxyl groups (Parks 1965, 1984)

$$M(OH)_n \leftrightarrow [M(OH)_{n-1}O]^- + H^+$$

As a result the cations are concentrated at the crack surfaces. In such a case the double electric layers are formed in the vicinity of the crack walls.

Rock deformation due to the tectonic stresses is accompanied by underground fluid migration. This migration is supposed to occur at the higher depth up to several kilometers (Nikolaevskiy 1966), and it can be especially intensive near the fault zone. Moving along the crack/channel, the fluid carries anions, and thus produces an extrinsic electric current. The electrokinetic current density averaged with respect to the cross section can be written as

$$j_e = -\sigma_r C \nabla P \qquad (5.1)$$

$$C = \frac{\varepsilon \varepsilon_0 \zeta}{\eta \sigma_f} \qquad (5.2)$$

where C is the streaming potential coefficient, σ_r is the average rock conductivity, σ_f is the fluid conductivity, ∇P is gradient of the fluid pressure in the cracks, ε and η are the dielectric permeability and viscosity of fluid, ε_0 is the electric constant and ζ denotes the potential difference across the electric double layer on crack walls (ζ-potential). Since the solid matrix conductivity is much smaller than that for the fluid the average rock conductivity in Eq. (5.1) is mainly determined by the fluid content. It should be noted that single pores and cracks can not be the conductor for the fluid flow as well as for the electrokinetic current (5.1) and thus only those cracks and channels, which create a connected system or cluster are able to contribute into the conductivity σ_r. In the non-conductive matrix approximation the rock conductivity is

$$\sigma_r \sim \sigma_0 (n - n_c)^\tau \qquad (5.3)$$

where n is the rock porosity, n_c is the percolation threshold, σ_0 is constant with dimension of conductivity and τ is the transport critical exponent (Feder 1988, Stauffer 1979). In what follows we shall use Eq. (5.3) in order to describe solely the rock conductivity due to formation of the percolation cluster filled by the water. In reality the rock conductivity is never zero because there is ion conductivity of the solid matrix, i.e. the percolation threshold of the rock conductivity is absent. In our case n_c denotes the percolation threshold for the electrokinetic current due to fluid flow.

An infinite cluster has a fractal structure above the percolation threshold within spatial scale, which does not exceed the correlation length

$$\xi \sim \frac{\xi_0}{\left|n - n_c\right|^\nu} \qquad (5.4)$$

where ν is the correlation length critical exponent, ξ_0 is constant of dimension of length. Once such fractal structure is formed in rock, its characteristic scale is of the order of the correlation length (5.4).

It is assumed that the upper crust includes great deals of small-scaled fluid-filled inhomogeneities, reservoirs with nonhydrostatic fluid, fracture zones and others. Some of such formation can be unstable, for example the sealed underground compartments with high pore pressure may become unstable by weak seismic events (Bernard 1992, Fenoglio et al. 1995). The typical size of the inhomogeneities can vary from several meters up to several kilometers. The focal zone of a forthcoming earthquake is frequently associated with such an unstable zone.

Consider an inhomogeneity, which includes high-permeability water-saturated rock. High pore pressure in this inhomogeneity is capable to sustain the outward fluid migration. Suppose that the pore space in this zone exhibits fractal structure. In this case the characteristic size of the inhomogeneity, L, is of the order of the correlation length (5.4), i.e. $L \sim \xi$. Combining the Eqs. (5.1) - (5.4) one can find a rough estimation of the electric current density due to the fluid migration

$$j_e \sim -\sigma_0 C \left(\frac{\xi_0}{L}\right)^{\frac{\tau}{\nu}} \nabla P \qquad (5.5)$$

The electrokinetic current (5.1) is partially compensated by the conduction current, which can flow in the fluid as well as in the solid matrix. Stationary current distribution in the rock obeys continuity equation $\nabla \cdot (\mathbf{J}_e + \sigma \mathbf{E}) = 0$, where E is electric field strength. The electric current is characterized by another average rock conductivity, σ that has no percolation threshold and thus does not coincide with σ_r. Introducing the electric potential through $\mathbf{E} = -\nabla\varphi$, we get

$$\sigma\nabla^2\varphi + (\nabla\sigma \cdot \nabla\varphi) = \nabla \cdot \mathbf{j}_e \qquad (5.6)$$

The term on the right-hand side of Eq. (5.6) plays a role of the source. Let V be the volume occupied by the electrokinetic currents so that $\mathbf{j}_e$ and C vanishes outside this zone. Assuming for the moment that σ is uniform, then at the large distance, when $r \gg L$, the solution of Poisson equation (5.6) can be written as

$$\varphi = \frac{\mathbf{d} \cdot \mathbf{r}}{4\pi\sigma r^3}, \quad \mathbf{d} = \int_V \mathbf{j}_e dV \qquad (5.7)$$

where d is the effective dipole moment of the electrokinetic currents $\mathbf{j}_e$ and $V \sim L^3$.

It should be noted that in the uniform medium, i.e., σ and C are constant everywhere, the conduction and electrokinetic currents are completely compensated by each other in all space including the volume V, so that the effective moment d becomes zero (Surkov 2000, Fedorov et al. 2001). Actually this compensation does not arise because the coefficients σ and C are different in the inhomogeneity and surrounding rock due to the change in porosity. Hence one obtain the rough estimation $|\mathbf{d}| \sim |\mathbf{j}_e| L^3$. Notice that this estimation is valid except for the case of the nonrealistic spherically symmetric current distribution when d equals zero.

We also assume that $|\nabla P| \sim \Delta P/L$, where ΔP is the pore fluid pressure difference between the inhomogeneity and surrounding rock. In the case of large-scaled inhomogeneities such as earthquake hypocenter, ΔP is supposed to be proportional shear stress drop, Δs, caused by rock fracture before main shock. The shear stress drop is of the order of crushing/shear strength and thus Δs is independent of the size L (Scholz 1990). We assume the same as to ΔP, i.e. ΔP is independent of the size L.

Taking into account these expressions and substituting Eq. (5.5) into Eq. (5.7) gives the potential φ. Finally we obtain the dependence of the electric field on distance r and inhomogeneity size L:

$$E \sim \frac{C\Delta P\sigma_0 L^2}{4\pi\sigma r^3}\left(\frac{\xi_0}{L}\right)^{\frac{\tau}{\nu}} \sim L^{2-\frac{\tau}{\nu}} \qquad (5.8)$$

Note that for the non-fractal inhomogeneity we get $E \sim L^2$.

In the model reported by Surkov et al. (2002) the permeability and porosity are maximal in the center of inhomogeneity and decrease in the radial directions so that the porosity gradient can be estimated as $d_r \sim \Delta n/L$, where L denotes the characteristic size of the inhomogeneity. The central non-fractal part/core of the inhomogeneity consists of broken rock with so high permeability and porosity that the correlation length tends to zero in this region. The core is surrounded by the layer $L - H < r < L$ with fractal structure, where the porosity decreases down to the percolation threshold n_c. The thickness, H, of this layer is estimated as $H \sim L^{v/(v+1)}$. In such a case the effective dipole moment of the electrokinetic currents in the fractal region is $|d| \sim |j_e| L^2 H$, where the electrokinetic current density is

$$|j_e| \sim \sigma_0 C \left(\frac{\xi_0}{H} \right)^{\frac{\tau}{v}} \frac{\Delta P}{L}$$

Hence

$$E \sim L^{1 - \frac{\tau - v}{v+1}} \tag{5.9}$$

The typical size of earthquake focal zone, L, can be related with the earthquake magnitude, M, by the empirical rule (Kanamori and Anderson 1975):

$$\log L = 0.5M - 1.9 \tag{5.10}$$

where L is measured in kilometers. Combining the Eq. (5.9) and (5.10), yields (Surkov et al. 2002)

$$\log E = aM + b, \qquad a = 0.5 \left(1 - \frac{\tau - v}{1 + v} \right) \tag{5.11}$$

where b is constant. Using the critical exponents $\tau = 1.6$, $v = 0.88$ obtained by numerical simulation on three-dimensional grids (Feder 1988, Stauffer 1979) gives $a \approx 0.31$. The same dependence has been reported by Varotsos et al. (1996) for the so-called seismic electric signals (SES) occasionally observed prior to strong crust earthquake occurrence. Interestingly enough the empirical factor 'a' presented by Varotsos et al. (1996) is $a \approx 0.34 - 0.37$, that is close to the above predicted value. One should note that for the non-fractal inhomogeneity $a = 1$.

5.4 Estimation of the signal duration

It is known that in conductive layers of the upper crust the low-frequency (ULF band) electromagnetic field and currents mainly propagate due to diffusion process. In such a case an electromagnetic field perturbation propagates as $r^2 \sim t/(\mu_0\sigma)$, where r is the distance from the source, σ is the rock basement conductivity and μ_0 denotes the magnetic constant. Hence the moment of the perturbations arrival at a fixed point r obeys the law: $t \sim \mu_0\sigma r^2$. Taking the characteristic distance $r = 100$ km and $\sigma = 10^{-3}$ S/m we obtain the time interval $t \sim 10$ s, that is much less than observed single SES duration (several tens minutes). Based on this estimation it is reasonable to assume that the SES duration is determined, in the first place, by the duration of mechanical processes in the source itself.

In order to evaluate the electromagnetic signal duration consider a highly permeable fractured area with higher fluid pressure surrounded by lower permeable rock or vise versa a lower permeable area surrounded by water-saturated rock. The changes of tectonic stresses can result in the rock fracture. Formation of fresh cracks and voids leads to the increase of permeability, and then fluid starts to flow from higher fluid pressure area to lower one. First we ignore the fractal properties of the pore space. Suppose that the fluid begins to move from the surrounding rock towards the center of the fracture zone/area. This process will be lasted until the pore fluid pressure in the fracture zone becomes equal to that in the surrounding space.

Consider a simplified model of the medium instead of study with a rigorous formulation of the problem. The fluid velocity, v, in narrow channel/crack is much smaller than the sound velocity C_f in the fluid. Therefore the fluid density variations, $\delta\rho$ is small, i.e. $\delta\rho<<\rho_0$, where ρ_0 is undisturbed fluid density, and the linearized equation for fluid motion can be used

$$\rho_0\partial_t v = -\nabla P - \frac{\eta v}{k} \tag{5.12}$$

where P is variations of the pore fluid overpressure (above hydrostatic pressure), η is the coefficient of the fluid viscosity and k is the rock permeability, $\partial_t = \partial/\partial t$ denotes partial time-derivative. Here we have ignored the fluid velocity distribution in the channel and solid matrix compressibility as well. In the stationary case, when $\partial_t v = 0$, Eq. (5.12) is converted to the form

$$v = -\frac{k}{\eta}\nabla P \qquad (5.13)$$

that coincides with well-known Darcy's law.

The linearized continuity equation is given by

$$\partial_t \delta\rho + \rho_0 \nabla \cdot v = 0 \qquad (5.14)$$

The fluid density variation, $\delta\rho$ can be related to the variation of the pore fluid overpressure, i.e. $P = C_f^2 \delta\rho$. Combining this expression with the Eqs. (5.12) and (5.14), yields

$$\partial_t^2 P = C_f^2 \nabla^2 P - \frac{\eta}{k\rho_0}\partial_t P \qquad (5.15)$$

where k is the average rock permeability and P makes sense of the fluid pressure averaged over elementary rock volume. In the low frequency case the second order time-derivative on the left-hand side of Eq. (5.15) can be omitted. Then we come to diffusion-type equation for the average pressure

$$\partial_t P = K\nabla^2 P \qquad (5.16)$$

Here K is the diffusion coefficient for the fluid moving along the underground channels

$$K = \frac{k\rho_0 C_f^2}{\eta} = \frac{kG_f}{\eta} \qquad (5.17)$$

where G_f is the compressibility modulus of the fluid. Note that, in spite of the simplified approach, our estimation of diffusion coefficient coincides with exact solution obtained by Frenkel (1944) for the more complicated model at least by the order-of-magnitude

$$K = \frac{kG_f}{\eta\alpha}, \quad \text{where} \quad \alpha = 1 + \left[\frac{1}{n}\left(1 - \frac{G}{G_s}\right) - 1\right]\frac{G_f}{G_s}$$

Here G_s and G are the compressibility modules of the solid matrix and the dry porous rock and n is porosity.

The diffusion duration can be roughly estimated from Eq. (5.16)

$$t \sim \frac{L^2}{K} \qquad (5.18)$$

where L is the characteristic size of the inhomogeneity/fracture zone. For example, taking the parameters $L = 0.3$ km, $k = 10^{-12} - 10^{-14}$ m^2, $\eta = 10^{-4}$ kg/

(m·s) and $G_f = 2.3 \times 10^9$ Pa one can find that $t \sim 1 - 100$ h. The duration of typical single SES as observed by Varotsos et al. (1996) is of the order of one hour.

Now we consider a model problem for infinite fracturing medium with fractal structure in order to estimate the duration of the fluid migration in fractal pore space. The medium contains a region/inhomogeneity with high pore fluid pressure. This inhomogeneity plays a role of natural reservoir or the fluid source. Once the percolation threshold is exceeded, say due to stress variation and energization of crack formation, the fluid begins to penetrate into surrounding rock. Suppose the fluid distribution is spherically symmetric on average so that all quantities depend on distance r from the origin.

It follows from Eq. (5.17) that the diffusion coefficient is proportional to the rock permeability k. According to the percolation theory, the permeability depends on the porosity in the same way as conductivity (5.3), i.e.

$$k \sim \left(n - n_c\right)^\tau \qquad (5.19)$$

where τ is the transport critical exponent (Feder 1988). This quantity, k, can be related to the active porosity, n_a, which includes only those channels and cracks that belong to percolation cluster

$$n_a \sim \left(n - n_c\right)^\beta \qquad (5.20)$$

where β is the order parameter critical exponent that can be expressed through the fractal dimension D of the percolation cluster: $\beta = v(3-D)$. The active porosity varies with distance in the fractal zone as r^{D-3}. Combining this dependence with the Eqs. (5.17), (5.19) and (5.20), gives

$$K(r) = \frac{K_0}{r^\theta} \qquad (5.21)$$

where K(r) is the coefficient of abnormal diffusion in fractal zone, $\theta = \tau/v$ is the critical exponent of the diffusion coefficient and K_0 is constant. It should be noted that in a fractal medium the pore fluid pressure P is highly nonanalytic function, and in fact contains singularities on all length scales. To simplify the problem, one can consider the smooth envelope of this function P(r, t), where r is the distance from the center of the inhomogeneity. It follows from the numerical modeling that the dynamics of P(r, t) are well approximated by the differential equation (O`Shaughnessy and Procaccia 1985a, b)

$$\partial_t P = \frac{1}{r^{D-1}} \partial_r \left(r^{D-1} K(r) \partial_r P \right) \tag{5.22}$$

where $\partial_r = \partial/\partial r$. Eq. (5.22) describes the diffusion on fractal with dimension D, in contrast to usual diffusion equation (5.16), which is valid as D = 3. In a word, Eq. (5.22) is a generalization of spherically symmetric diffusion equation in three dimensional spaces.

Let a be the radius of the inhomogeneity/source. Multiplying both sides of Eq. (5.22) by the factor r^{D-1} and integrating with respect to r from a to infinity, gives

$$\int_a^\infty r^{D-1} \partial_t P(r, t) dr = -K_0 a^{D-\theta-1} \partial_r P(a, t) \tag{5.23}$$

Here we have taken into account that $r^{D-\theta-1} \partial_r P \to 0$ when $r \to \infty$. Let m (t) be the net fluid mass penetrated into the surrounding rock for the time interval t. The fluid mass crossing the surface confined by the radius r = a per unit of time is $d_t m(t) = \rho_0 v(a, t) S$, where $S \sim a^{D-1}$ denotes the surface area in space with dimension D. The fluid velocity, v (a, t) obeys the Darcy law (5.13), where the rock permeability k is given by Eq. (5.19). As before we find that $k \sim r^{-\theta}$ and hence

$$d_t m(t) \sim -\frac{\rho_0}{\eta} a^{D-\theta-1} \partial_r P(a, t) \tag{5.24}$$

Combining Eq. (5.23) and (5.24), gives the boundary condition at r = a. For simplicity consider the case of point source when $a \to 0$

$$\int_0^\infty r^{D-1} P(r, t) dr = \alpha m(t) \tag{5.25}$$

Here m(t) is given function and α is constant. The solution of equation (5.22) with the condition (5.25) has the form

$$P(r, t) = \frac{\alpha}{\Gamma\left(\frac{D}{2+\theta}\right) K_0^{\frac{D}{2+\theta}} (2+\theta)^{\frac{2D}{2+\theta}-1}} \int_0^t \exp\left(-\frac{r^{2+\theta}}{K_0(2+\theta)^2(t-t')}\right) \frac{d_t m(t') dt'}{(t-t')^{\frac{D}{2+\theta}}} \tag{5.26}$$

where $\Gamma(x)$ denotes Γ-function. It follows from Eq. (5.26) that the integral depends on the dimensionless parameter $r^{2+\theta}/\left[K_0(2+\theta)^2 t\right]$. Once L is the characteristic spatial scale of the region, the time of fluid migration into this region has the form

$$t = L^{2+\theta}/\left[K_0(2+\theta)^2\right] \sim L^{2+\theta} \tag{5.27}$$

that is typical for a random walker on fractal medium. If the quantity (5.27) is much greater than the characteristic period of function d_tm (t), one may replace it by the delta-function and take the integral in Eq. (5.26). In such a case Eq. (5.27) defines the characteristic duration of both the fluid diffusion and the single SES.

For the more complicated model of inhomogeneity (Surkov et al. 2002), where the high permeable core is surrounded by fractal zone with spatial scale $H \sim L^{v/(v+1)}$, the duration of diffusion obeys

$$t \sim H^{2+\theta} \sim L^{v(2+\theta)/(1+v)} \tag{5.28}$$

Taking a notice of the fact that this model gives a good correlation with the SES observation before earthquakes, at least as for the SES amplitude (5.10), we shall try to estimate the SES duration in the same way. Combining Eqs. (5.10) and (5.28) we find

$$\log t = a_1 M + b_1, \qquad a_1 = \frac{v(2+\theta)}{2(1+v)} \tag{5.29}$$

where M is the magnitude of impending earthquake and b_1 is constant. Taking above parameters for the critical exponents v and θ, gives $a_1 \approx 0.89$. We can not compare the predicted dependence (5.29) with observation because proper statistics of the single SES duration is still absent. One should note that for the non-fractal inhomogeneity $a_1 = 1$.

5.5 Discussion

The scaling analyses presented in this chapter reveal that the fractal properties of pore space considerably affect the electrokinetic effect. The amplitude and duration of the seismic electric signals originated from the electrokinetic effect depend on both the correlation length and transport critical exponents. These estimations can be applied to the near and pre-seismic electromagnetic phenomena possibly associated with forthcoming earthquakes. In particular we have related the SES parameters with earthquake magnitude. It was found that in the case of fractal medium the coefficient 'a' that relates $\log E$ to M becomes about one-third compared to that in the case of non-fractal medium. It is worth mentioning that such estimations is sensitive to topology of the real pore space, structure of the water-saturated region and the fluid pore pressure gradient. For example, two above considered models, i.e. the fractal inhomogeneity above the percolation threshold and the fractured region with high permeable core surrounded by the fractal zone, reveal different relationships between the SES parameters, the fractal zone size and earthquake magnitude (cf. the Eqs. (5.8) and (5.9) or the Eqs. (5.27) and (5.28)).

Another mechanism for the ULF electromagnetic field variations observed before strong crust earthquake have been recently proposed by Surkov (1997; 1999; 2000) and Surkov et al. (2003). The electromagnetic noises can be excited by acoustic emissions associated with crack formation at the Earth conductive stratums immersed in a uniform geomagnetic field. Energization of crack formation and fracture process in a focal zone before the main shock gives rise to the electromagnetic variations, whose amplitudes are proportional L^3. Hence, as it follows from Eq. (5.10), $\log E \sim 1.5M$ in contrast to the Eqs. (5.27) and (5.28). This fact makes it possible to distinguish these different mechanisms for telluric electric signals.

The SES duration was estimated in terms of the time needed for the fluid to penetrate from the fractured fractal region with high fluid pressure into the surrounding rock or vise versa depending on the model. It should be noted, this estimation is based on the assumption that percolation cluster are formed for short time compared to the fluid migration. Besides we have omitted such details as possible formation of the "viscous fingers" at the front of fluid penetrating in the rock and the anisotropy of the rock permeability. Once the anisotropic disordered medium are formed, the fluid filtration with a fixed bias direction, say due to gravity force, can arise. As the porosity is far above the percolation threshold the fluid motion can exhibit log-periodic oscillations in the effective exponent versus time (Stauffer and Sornette 1998, Bustingorry and Reyes 2000). Note that this effect has analogy with diffusion of low-frequency electromagnetic field into magnetized plasma when anisotropic plasma conductivity takes place (Surkov 1996). One may expect appearance of the SES oscillations due to such effect.

Acknowledgments

This research was partially supported by ISTC under Research Grant No. 1121.

5.6 References

Bernard P (1992) Plausibility of long electrotelluric precursors to earthquakes. J Geophys Res 97: 17531-17546

Bustingorry S, Reyes ER (2000) Biased diffusion in anisotropic disordered systems, Phys Rev E 62: 7664-7669

Fedorov EN, Pilipenko VA, Vellante M, Uyeda S (2001) Electric and magnetic fields generated by electrokinetic processes in a conductive crust. Phys Chem Earth C 26:793-799

Fenoglio MA, Johnston MJS, Byerlee JD (1995) Magnetic and electric fields associated with changes in high pore pressure in fault zones: Application to the Loma Prieta ULF emissions. J Geophys Res 100:12.951-12.958

Feder E (1988) Fractals. Springer, Berlin Heidelberg New York

Frenkel Ya I (1944) On the theory of seismic and seismoelectric phenomena in moist soil. Proc Academy of Sciences USSR series of geography and geophysics 8: 133-150

Ivanov AG (1940) Seismoelectric effect of the second kind. Proc Academy of Sciences USSR series of geography and geophysics 4: 699-726

Jouniaux L, Pozzi JP (1999) Streaming potential measurements in laboratory: a precursory measurement of the rupture and anomalous 0.1- 0.5 Hz measurements under geochemical changes. In: Hayakawa M(ed) Atmospheric and ionospheric phenomena associated with earthquakes TERRAPUB, Tokyo, pp. 873-880

Kanamori H, Anderson DL (1975) Theoretical basis of some empirical relations in seismology. Bull Seism Soc Amer 65:1073-1095

Martner ST, Sparks NR (1959) The electroseismic effect. Geophysics 24: 297-308

Mizutani H, Ishido T (1976) A new interpretation of magnetic field variation associated with the Matsushita earthquakes. J Geomagn Geoelectr 28:179-188

Nikolaevskiy VN (1996) Geomechanics and Fluidodynamics. Dordecht Boston, London

Parks GA (1965) The isoelectric points of solid oxides, solid hydroxides, and aqueous hydroxo complex systems. Chem Rev 65:177-198

Parks GA (1984) Surface and interfacial free energies of quartz. J. Geophys. Res. 89:3997-4008

O`Shaughnessy B, Procaccia I (1985a) Analytical solutions for the diffusion on fractal objects. Phys Rev Lett 54:455-458

O`Shaughnessy B, Procaccia I (1985b) Diffusion on fractals. Phys Rev 32A: 3073-2083

Scholz CH (1990) The Mechanics of earthquakes and faulting. Cambridge Univ. Press, Cambridge UK

Stauffer D, Sornette D (1998) Log-periodic oscillations for biased diffusion on random lattice. Physica A 252: 271-277

Surkov VV (1996) Front structure of the Alfven wave radiated into the magnetosphere due to excitation of the ionospheric E layer. J Geophys Res 101:15403-
15409

Surkov VV (1997) The nature of electromagnetic forerunners of earthquakes,
Trans (Doklady) Russian Acad Sci. Earth Science Sections 355:945-947

Surkov VV (1999) ULF electromagnetic perturbations resulting from the fracture
and dilatancy in the earthquake preparation zone. In: Hayakawa M (ed) Atmospheric and Ionospheric Phenomena Associated with Earthquakes
TERRAPUB, Tokyo, pp. 357-370

Surkov VV (2000) Electromagnetic effects caused by earthquakes and explosions.
MEPhI, Moscow

Surkov VV, Uyeda S, Tanaka H, Hayakawa M (2002) Fractal properties of medium and seismoelectric phenomena. J Geodynamics 33: 477-487

Surkov VV, Molchanov OA, Hayakawa M (2003) Pre-earthquake ULF electromagnetic perturbations as a result of inductive seismomagnetic phenomena
during microfracturing. J Atmosphere and Solar-Terrestrial Physics 65:31-46

Uyeda S, Nagao T, Orihara Y, Yamaguchi T, Takahashi I (2000) Geoelectric potential changes: Possible precursors to earthquakes in Japan. Proc Natl Acad
Sci 97: 4561-4566

Varotsos P, Alexopoulos K (1984a) Physical properties of the variations of the
electric field of the earth preceding earthquakes, I. Tectonophysics 110:73-98

Varotsos P, Alexopoulos K (1984b) Physical properties of the variations of the
electric field of the earth preceding earthquakes, II Determination of epicenter
and magnitude. Tectonophysics 110:99-125

Varotsos P, Lazaridou M, Eftaxias K., Antonopoulos G, Makris J, Kopanas J
(1996) Short term earthquake prediction in Greece by seismic electric signals.
In: Lighthill S J (ed) A Critical Review of VAN, World Scientific, Singapore.
pp 29-76

Chapter 6. Fractal Network and Mixture Models for Elastic and Electrical Properties of Porous Rock

M. Pervukhina, Y. Kuwahara, H. Ito

Geological Survey of Japan, Higashi, Tsukuba, Ibaraki, 305-8567 Japan.

6.1 Summary

Study of the correlation of elastic and electrical properties of porous rock is important for predicting parameters, such as porosity and permeability. We review the methods that allow calculations of both electrical and elastic properties of porous rock for the same microstructure. The methods are categorized into mixture theories and fractal networks theories. The mixture theories describe electrical and elastic properties for limited cases of microstructures and fail to predict properties of porous rock near percolation. Fractal models, recently developed for elastic and electrical properties of porous rock are described for conductivity and seismic velocities for a wide range of isolated and interconnected pore geometries. We review the application of the model to the results of collocated seismic velocity tomography and magnetotelluric experiments at an active faults of the experimental data using the fractal model illustrated that the deep extension of the active fault can be regarded as the region, with small aspect ratios equal to 10^{-2}, showing that the pore geometry in the region is far from the interfacial energy-controlled fluid geometry.

6.2 Introduction

Theoretical problems of effective properties of elasticity and resistivity of composite material have been intensively studied for a long time in the field of material science. This kind of problem has also become of practical importance in the field of solid earth science. Examples of the problem are found in collocated magnetotelluric (MT), seismic velocity tomography studies of the crust and the upper mantle and in borehole measurements of resistivity and seismic velocities. The investigation of the correlations of elastic and electrical properties of porous rock is important for understanding of the structure of studied area and for estimation of non-measured parameters, such as porosity and permeability. The precise nature of porous rock analysis of both elastic and electromagnetic data is very interesting, especially for percolation. Experimental data of electrical resistance and seismic velocities must be analyzed using the method that is able to calculate both properties for the same microstructure. However, most of interpretations of the experimental data are based on inconsistent theories and/or empirical dependencies. This situation is possibly caused from the fact that the conventional mixture theories (Berryman 2000) describe electrical and elastic properties only for limited cases of microstructures and fail to predict properties of porous rock near percolation.

Network theories are other plausible approach to describe rock that is mixture of crystals, voids, cracks, and fractures. Methods based on random networks, for example, renormalization group method (Madden 1983), percolation theory (Gueguen et al. 1991) and fractal random networks (Bahr 1997) are a standard approach for calculation of transport properties, namely, conductivity and permeability of rock. Spangenberg (1998) developed a fractal model for elastic properties of porous rock. Pervukhina et al. (2003) modified his model to describe isolated pore case. Application of the modified model to the electrical properties allowed calculation of seismic velocities and resistivities for a wide range of microstructures including 3D grain and pore anisotropy and different interconnection extent from isolated to interconnected pores.

In this chapter, we review the theories for elastic and electrical properties of porous medium. These theories are conventionally divided into mixture theories and network theories as shown in Fig.6.1. We examine the mixture models for elastic and electrical properties of porous rock, paying particular attention to the bounds and estimates, which could be made for both electrical and elastic properties for the same microstructure. Then fractal network models allowing calculation of both elastic and electrical properties are discussed. Finally, we review an application of the fractal model to results of a collocated seismic velocity tomography and magnetotelluric experiments.

6.3 Mixture theories

Mixture theories are used to determine effective properties of composites. The theories do not consider reflection and refraction of the electrical and elastic wave on inhomogeneity, assuming that the size of inhomogeneity is much smaller than the length of elastic wave in the case of elastic properties or the free path length in the case of electric properties (so-called, long-wave approximation). We restrict our review with electrical and elastic properties of two-phase mixtures, which are commonly applied for calculation of physical properties of rocks.

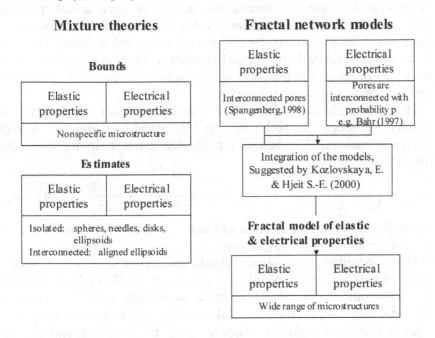

Fig. 6.1 Mixture theories and network models of electrical and elastic properties of porous rocks

There are two types of results obtained in the theory of mixtures: bounds obtained for nonspecific geometry of constituents and results obtained for a specific microstructure of composite. Hereafter we call these as bounds and estimates, respectively. Bounds are based on thermodynamic stability criteria or on variational principles and can be calculated for both electrical and elastic properties of mixtures. The theoretical results obtained for a specific mixture microstructure are supposed to satisfy the bounds and play essential role in evaluating of the rock properties.

6.3.1 Bounds

Voigt (1910) and Reuss (1929) determined bounds for highly anisotropic composite materials. The assumption of homogeneous strain (Voigt 1910) results in parallel connection of equivalent network or arithmetic mean of elastic moduli. The assumption of homogeneous stress (Reuss 1929) leads to a serial connection or a harmonic mean of elastic moduli of the composures. Using variational principles, Hill (1952) showed that the assumptions of homogeneous strain and strain are the rigid upper and lower bounds for all possible mixture microstructures including anisotropic ones. As regards to electrical properties, the bounds of anisotropic mixtures are the resistivities of horizontally and vertically layered medium. The resistivity of the horizontal and vertical lamellar medium can be calculated as resistivity of equivalent networks as arithmetic mean for serial connections and as harmonic mean for parallel ones.

Hashin and Shtrikman (1961; 1962) used a variational approach to determine the bounds for both elastic and electrical properties for isotropic composite materials. The bounds for electrical conductivity for solid-liquid mixture can written as

$$\sigma_{HS^-} \equiv \left(\frac{\beta}{3\sigma_m} + \frac{1-\beta}{\sigma_p + 2\sigma_m} \right)^{-1} - 2\sigma_m \le \sigma_{eff} \le \left(\frac{\beta}{\sigma_m + 2\sigma_p} + \frac{1-\beta}{3\sigma_p} \right)^{-1} - 2\sigma_p \equiv \sigma_{HS^+}$$

where β is a volumetric fraction of fluid and σ_m and σ_p are conductivities of matrix and pore fluid, respectively.

Hashin-Shtrikman bounds for elastic moduli are usually presented in terms of bulk and shear moduli and sometimes for Young's modulus and Poisson's ratio. In the present article we mainly use elastic modulus $M=V_p^2\rho$, where V_p is compressional wave velocity and ρ is density. Considering that the modulus M equals to $M=K+4/3\mu$, where K and μ is bulk and shear moduli, respectively, we rewrite the Hashin-Shtrikman bounds for the modulus M as

$$M_{HS^-} \equiv \left(\frac{\beta}{M_p} + \frac{1-\beta}{M_m + \frac{4}{3}\mu_m} \right)^{-1} \le M_{eff} \le \left(\frac{\beta}{M_p + \frac{4}{3}\mu_m} + \frac{1-\beta}{M_m} \right)^{-1} - \frac{4}{3}\mu_m \equiv M_{HS^+}$$

Here M_m, μ_m, M_p and μ_p are elastic moduli for matrix and pore materials respectively.

Hashin-Shtrikman bounds for shear modulus are

$$\mu_{HS^-} \equiv 0 \le \mu_{eff} \le \frac{\mu_m}{2} \left[\left(\frac{\beta}{\varsigma} + \frac{(1-\beta)}{2+\varsigma} \right)^{-1} - \varsigma \right] \equiv \mu_{HS^+}$$

Here $\varsigma = \dfrac{9M_m - 4\mu_m}{3M_m + 2\mu_m}$.

Hashin-Shtrikman's upper and lower bounds are the best ones that could be done for elastic and electrical properties of isotropic mixture without a priori knowledge of composite microstructure. The useful alternative is to specify geometry of the phases to get more accurate estimation of electrical and elastic properties of mixture. Below the electric and elastic properties are estimated for some particular geometry of phases.

6.3.2 Estimates

Among the numerous estimates for effective properties of porous rock, we here review the estimate theories that can be made for elastic and electrical properties for the same microstructure (Fig. 6.1). Firstly, we refer several estimates for both the properties for homogeneous media with isolated inclusions of different simple shapes, namely, spheres, needles, disk and axis symmetric ellipsoids. It is noted that these results are valid for an assumption of non-interactions between the inclusions, i.e. for dilute concentrations of fluid inclusions. Secondly, we mention an estimate for elastic and transport properties of two-phase medium with isotropic matrix and aligned interconnected ellipsoidal inclusions (Bayuk and Chesnokov 1998). This estimate is considering high concentration of inclusions and taking into account the microstructure of the porous medium.

Isolated inclusions

For clarity of presentation we refer unified formulas presented by Berryman (2000), who generalized diverse results for effective electrical and elastic properties of isotropic composite with isolated inclusions of different shapes.

Effective electrical conductivity σ_{eff} of inclusions of different shapes in an isotropic matrix was obtained by Cohen et al. (1973) and Galeener (1971). For two-phase mixtures it can be calculated as

$$\frac{\sigma_{eff} - \sigma_m}{\sigma_{eff} + 2\sigma_m} = \beta(\sigma_p - \sigma_m)f ,$$

where σ_m, σ_p are the conductivity of matrix and of pore filling and the factor f is presented in Table 6.1 for different shapes of inclusions. The formula for isolated axis symmetric ellipsoidal conductive inclusions was derived by Semjonow (1948) as:

$$\sigma_{eff} = \frac{1}{3}\left(\sigma_1 + \sigma_2 + \sigma_3\right)$$

where

$$\sigma_i = \sigma_m \frac{(1-\beta)(n_i - 1)\sigma_m + (n_i - (n_i - 1)(1-\beta))\sigma_p}{(n_i - 1 + \beta)\sigma_m + (1-\beta)\sigma_p} , i = 1, 2, 3 ,$$

$$n_1 = n_2 = \frac{-2h^3}{h - (1 + h^2)\tan^{-1} h},$$

$$n_3 = \frac{-h^3}{(1 + h^2)(\tan^{-1} h - h)},$$

$$h = \sqrt{\alpha^{-2} - 1},$$

where α is the aspect ratio of the axis symmetric inclusions.

Table 6.1 Three examples of coefficients for different shape of inclusions in isotropic composites.

Inclusion shape	Factor f
Spheres	$\dfrac{1}{\sigma_p + 2\sigma_m}$
Needles	$\dfrac{1}{9}\left(\dfrac{1}{\sigma_m} + \dfrac{4}{\sigma_p + \sigma_m}\right)$
Disks	$\dfrac{1}{9}\left(\dfrac{1}{\sigma_p} + \dfrac{2}{\sigma_m}\right)$

Kuster and Toksöz (1974) derived estimates of bulk and shear moduli of composites with isotropic matrix and isolated spherical inclusions. Theirs results were generalized by Wu (1966), Walpole (1969) and Walsh (1969) for the nonspherical inclusions as:

$$(K_{eff} - K_m)\frac{K_m + \frac{3}{4}\mu_m}{K_{eff} + \frac{4}{3}\mu_m} = \beta(K_p - K_m)P$$

and

$$(\mu_{eff} - \mu_m)\frac{\mu_m + \varsigma_m}{\mu_{eff} + \varsigma_m} = \beta(\mu_p - \mu_m)Q$$

Coefficients P and Q are presented in Table 6.2 for the same four type of inclusions, which were mentioned for electrical properties. Special characters used in Table 6.2 are defined by $\xi = \mu[(3K + \mu)/(3K + 4\mu)]$, $\gamma = \mu[(3K + \mu)/(3K + 7\mu)]$, and $\varsigma = (\mu/6)[(9K + 8\mu)/(K + 2\mu)]$. The expressions for penny-shaped cracks were derived by Walsh (1969) and assuming $K_p / K_m \ll 1$ and $\mu_p / \mu_m \ll 1$.

Table 6.2 The coefficients P and Q for different shapes of inclusions

Inclusion shape	P	Q
Spheres	$\dfrac{K_m + \dfrac{4}{3}\mu_m}{K_p + \dfrac{4}{3}\mu_m}$	$\dfrac{\mu_m + \varsigma_m}{\mu_p + \varsigma_m}$
Needles	$\dfrac{K_m + \dfrac{4}{3}\mu_m + \dfrac{1}{3}\mu_p}{K_p + \dfrac{4}{3}\mu_m + \dfrac{1}{3}\mu_p}$	$\dfrac{1}{5}\left(\dfrac{4\mu_m}{\mu_m + \mu_p} + 2\dfrac{\mu_m + \gamma_m}{\mu_p + \gamma_m} + \dfrac{K_p + \dfrac{4}{3}\mu_m}{K_p + \mu_m + \dfrac{1}{3}\mu_p} \right)$
Disks	$\dfrac{K_m + \dfrac{4}{3}\mu_p}{K_p + \dfrac{4}{3}\mu_p}$	$\dfrac{\mu_m + \varsigma_p}{\mu_p + \varsigma_p}$
Penny cracks	$\dfrac{K_m + \dfrac{4}{3}\mu_p}{K_p + \dfrac{4}{3}\mu_p + \pi\alpha\xi_m}$	$\dfrac{1}{5}\left(1 + \dfrac{8\mu_m}{4\mu_p + \pi\alpha(\mu_m + 2\xi_m)} + 2\dfrac{K_p + \dfrac{2}{3}(\mu_p + \mu_m)}{K_p + \dfrac{4}{3}\mu_m + \pi\alpha\xi_m} \right)$

Interconnected aligned ellipsoidal inclusions

Using the general singular approximation (GSA) method (Schermergor, 1977), Bayuk and Chesnokov (1998) suggested a way to calculate either elastic or electrical properties of rock for the single microstructure. This method allows considering high concentration of inclusions and takes into account the inner structure of the porous medium. The GSA method based on the assumption that the properties of composite body can be expressed using the effective operator $\mathbf{L}^{eff}$, which connects the average vector $<U>$ with average divergence of stress or electrical conductivity tensor $<LU>$ as

$$<LU> = \mathbf{L}^{eff}<U>$$

In the elasticity case, the operator $\mathbf{L}$ has the form $L_{ik} = \nabla_j C_{ijkl} \nabla_l$, where $\mathbf{C}$ is the fourth-rank tensor of the elastic constants and $\mathbf{U}$ is a vector of displacements. In case of electrical properties $L_i = T_{ij}\nabla_j$, here $\mathbf{T}$ is the two-rank tensor of electrical conductivity. Here we used the notation $\nabla_i = \dfrac{\partial}{\partial x_i}$.

The effective properties of the composite body are expressed through the properties of homogeneous body of the same size and shape, so-called comparison body, with the known tensor of elastic constants $\mathbf{C}^C$ and tensor of electrical conductivity $\mathbf{T}^C$. After algebraic transformation and neglecting the formal component of the second derivative of the Green function, Shermergor (1977) derived the formula of the effective properties of the inhomogeneous body as

$$X^* = \left\langle X(I - g(X - X^C))^{-1}\right\rangle\left\langle I - g(X - X^C))^{-1}\right\rangle^{-1},$$

where $\mathbf{X}=\mathbf{C}$ (4th-rank tensor) in the elasticity case and $\mathbf{X}=\mathbf{T}$ (2nd-rank tensor) in the electrical properties case. The tensor $\mathbf{g}$ is the singular part of the second derivate of the Green function. For ellipsoidal inclusions, the tensor g in the case of elasticity takes the form

$$g_{ijkl} = -\frac{1}{4\pi}\int n_{k)(j}\Lambda_{i)(l}^{-1}d\Omega,$$

$$\Lambda_{ij} = C_{ijkl}^c n_k n_l,$$

$$n_{k)(j}\Lambda_{i)(l}^{-1} \equiv \frac{1}{4}\left(n_{kj}\Lambda_{jl}^{-1} + n_{ki}\Lambda_{jl}^{-1} + n_{lj}\Lambda_{ik}^{-1} + n_{li}\Lambda_{jk}^{-1}\right).$$

In the case of electric properties,

$$g_{ij} = -\frac{1}{4\pi}\int n_{ij}\Lambda^{-1}d\Omega,$$

$$\Lambda = T_{ij}n_i n_j,$$

where $n_1 = \frac{1}{a_1}\sin\theta\cos\varphi$; $n_2 = \frac{1}{a_2}\sin\theta\sin\varphi$; $n_3 = \frac{1}{a_3}\cos\theta$,   and

$d\Omega = \sin\theta\partial\theta\partial\varphi$, wherein a_1, a_2 and a_3 are semi-axes of the ellipsoidal inclusions.

The properties of comparison body is usually chosen as

$$X^C = (1 - k\beta)X^m + kX^p,$$

where β is volumetric fraction of inclusions, $\mathbf{X}^m$ and $\mathbf{X}^p$ are tensors of elastic or electrical properties of matrix and pore material respectively, and k is percolation factor with numerical values in the interval [0,1].

6.4 Fractal network models

6.4.1 Elastic properties

Spangenberg (1998) proposed to describe a real porous rock as a combination of matrix, pore canals and contact region (Fig.6.2). The geometrical unit corresponds to cuboids of free eligible ratios of the edges l_1, l_2, l_3. Pore canals are described by canals with free eligible sizes a_1, a_2, a_3. Contact region can be filled by an arbitrary number of self-similar generations. Porosity and density can be calculated for this model with arbitrary number of substructures filling contact region. To calculate elastic properties of the model, Spangenberg (1998) subdivided the geometrical model into rectangular components of matrix material, contact region and pore fill, combined the components in serial or parallel connections and calculated the physical properties for the resulting "equivalent network". He suggested that the combination of serial and parallel connections for the elastic moduli corresponds to an assumption of regions of homogeneous stress (Reuss 1929) and regions of homogeneous strain (Voigt 1910), respectively. Accordingly, he fulfilled horizontal subdivision and vertical subdivision of the model. Then he showed that the principal dependencies on geometry, porosity, contact and composition are the same for the models corresponding to both kinds of subdivision and used for further calculations only the formulas corresponding to vertical subdivision of the model:

$$M_i^v = \frac{a_j a_k M_p}{l_j l_k} + \frac{l_i M_p M_c \left((l_j - a_j)a_k + (l_k - a_k)a_j \right)}{l_j l_k \left(a_i M_c + (l_i - a_i)M_p \right)} + \frac{(l_j - a_j)(l_k - a_k)l_i M_c M_m}{l_j l_k \left((l_i - a_i)M_c + a_i M_m \right)}$$

$$\mu_i^v = \frac{(l_j - a_j)(l_k - a_k)l_i \mu_c \mu_m}{l_j l_k \left((l_i - a_i)\mu_c + a_i \mu_m \right)}$$

where $M=Vp^2\rho$, Vp is compressional velocity and ρ is density, Mm, Mp, Mc are the moduli for matrix, pore filling and contact region respectively and μ_m, μ_p, μ_c are shear moduli for matrix, pore filling and contact area respectively. If the contact region is filled with N generation of self-similar substructures, the calculation of the elastic moduli is an iterative process. The M_c and μ_c for the smallest generation is assumed to be equal to M_m and μ_m respectively. Elastic moduli of the contact region for other generations can be obtained using above formula.

The compressional and shear seismic velocities can be calculated from the elastic moduli and density as below:

$$V_p = \sqrt{\frac{M^v}{\rho}}, \quad V_s = \sqrt{\frac{\mu^v}{\rho}}.$$

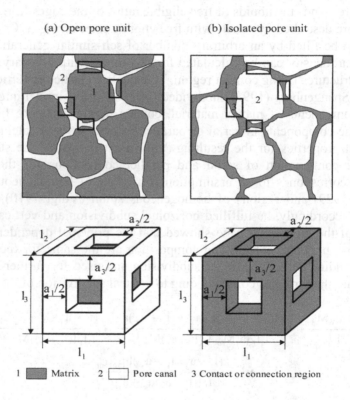

(a) Open pore unit   (b) Isolated pore unit

1 [Matrix]   2 [Pore canal]   3 Contact or connection region

Fig. 6.2 Open and isolated pore geometrical units. Contact regions of the open pore unit and connection regions of the isolated pore unit can be filled with N sub-generations of the self-similar units

6.4.2 Electrical properties

Numerous models were developed for the description of the electrical properties of binary mixtures of insulating matrix and conducting pores that are a very common case in rock physics (eg. Madden (1976; 1983) and Bahr (1997)). Random networks are investigated as models of such binary mixtures. The critical and averaging behavior of such network was investigated using different approaches, for example, numerical experiments and renormalization group method. These results were applied to a study of electrical conductivity of porous media. Kozlovskaya and Hjeit (2000) made an attempt to combine Bahr's model for electrical properties of porous rock and Spangenberg's model for elastic properties, so we consider Bahr's model in more details.

Bahr (1997) formulated the fractal network model that is a combination of the mixture theory with the percolation theory. He considered Hashin-Shtrikman upper bound under the assumption that volumetric fraction of high-conductive phase $\beta << 1$ and conductivity of pore filling is much higher than matrix one $\sigma_p >> \sigma_m$. He obtained the approximation derived by Waff (1974) for the conduction of the perfectly interconnected high-conductive phase. Then he introduced a dimensionless parameter C(p), so called "electrical connectivity", which depends on the probability p of high-conductive phase to be interconnected at any point of the medium. The introduction of C(p) extends Waff's approximation to the case of lower interconnections of high-conductive phase:

$$\sigma_{eff} = \frac{2}{3}\beta\sigma_m C(p).$$

Numerical values for C(p) are in the interval [0,1], where 1 stands for a perfectly interconnected high-conductive phase and 0 for 'isolated pockets' case.

Kozlovskaya and Hjeit (2000) suggested a simple link to connect the models and declared that the joint model describes the electrical and elastic properties for the same pore geometry. They assigned the connectivity $C_i(p)$ to be equal to

$$C_i(p) = 1 - \chi_i,$$

where i defines a direction of current propagation (i=1, 2, 3) and χ_i is the contact parameter introduced by Spangenberg and equals to

$$\chi_i = \prod_{g=0}^{N} \frac{(l_{j,g} - a_{j,g})(l_{k,g} - a_{k,g})}{l_{j,g} l_{k,g}},$$

where i, j, k=1, 2, 3 and $i \neq j \neq k$. N is the total number of generations of the model.

Such definition of C(p) does not allow description of effects near the percolation, because the Spangenberg's model describes interconnected pores. To overcome the above limitation the fractal model of elastic and electrical properties of porous rock was developed.

6.4.3 Elastic and electrical properties for a single microstructure

The fractal model was developed to describe both elastic and electrical properties of porous rock for the single microstructure by Pervukhina et al. (2003). The iterative process suggested by Spangenberg was extended to calculate electrical properties of the model with N sub-generations of self-similar units in the contact region.

The original model of Spangenberg (Fig. 6.2a) is failed to describe the electrical properties of pore rock with isolated pores. This impossibility results from the pore channel geometry. The geometry was designed not to describe the isolated inclusions case, but to describe the grains surrounded by water. To modify the model for describing the isolated pores case, the inverted geometrical unit with pore inclusion surrounded by matrix material was developed (Fig. 6.2b). The contact area in the original model, can be filled with N generations of the self-similar structures and determine the degree of pore interconnection of the model. Hereafter we will call the original model as open pore model and the inverted geometrical unit as the isolated pore model. The elastic moduli of the isolated pore model can be calculated similarly to the ones of the open pore model.

To calculate electrical properties of the unit, the following steps were taken: the geometric model is subdivided into rectangular areas of matrix material, pore filling and contact region, then the components were combined in a serial or parallel connection, and finally the electrical properties of equivalent network were calculated. Formulae for electrical resistivity and elastic moduli for open and isolated pore models are presented below. Spangenberg (1998) showed that the principal dependencies for both horizontal and vertical subdivision are the same, thus formulae for vertical subdivision are presented. Calculation of the electrical resistance of the model with N generation of self-similar models filling the contact region is iterative process. Firstly, the resistance of the smallest generation is calculated under the assumption that $r_c = r_m$ for the smallest generation. Then using the formula $r^i = R^i l_1^i l_2^i / l_3^i$, the resistivity of the contact region of the upper generation can be found, etc. The formulas for the resistance and the elastic moduli for both open and isolated pore model are presented in Table 6.3.

The conductivity for the vertical subdivision of Spangenberg model with 0 generation for open pore unit is presented in Fig. 6.3. The parameters for calculations are $\sigma_m = 10^{-4}$ S·/m and $\sigma_p = 10^{-1}$ S·/m; and geometric parameters of model: $l_1 = l_2 = l_3 = 1$ mm, $a_1 = a_2 = a_3 = 0...1$ mm. For comparison with Spangenberg mode, the results for other models, namely, Hashin-Strikman upper and lower bounds (HS+ and HS-), Waff's (1974) results for regularly arranged solid cubs surrounded by fluid and conductivity of a system containing fluid filled tubes along cubic grain edges (Grant and West 1965) are also shown in Fig 6.3. The obtained conductivity for the vertical subdivision of open pore model is in a good agreement with the results of the tubes (Grant and West 1965) for porosities less than 10%. The conductivities of the model throughout all porosities are enclosed between the curve for a system containing fluid filled tubes along grain edges (Grand and West 1965) and the curve for films along grain faces (Waff 1974).

Table 6.3 Resistance and elastic moduli for the vertical subdivision for both open and isolated pore models

	Interconnected pore model	Isolated pore model
R	$R = \dfrac{R_I R_{II}}{R_I + R_{II}}$, where $R_I = \dfrac{1}{a_1} \times$ $\dfrac{r_p l_3 (r_p a_3 + r_c (l_3 - a_3))}{r_p l_3 (l_2 - a_2) + r_p a_2 a_3 + r_c a_2 (l_3 - a_3)}$ and $R_{II} = \dfrac{1}{(l_1 - a_1)} \times$ $\dfrac{(r_p a_3 + r_c (l_3 - a_3))(r_c a_3 + r_m (l_3 - a_3))}{(r_c a_3 + r_m (l_3 - a_3)) a_2 + (r_p a_3 + r_c (l_3 - a_3))(l_2 - a_2)}$	$R = \dfrac{R_I R_{II}}{R_I + R_{II}}$, where $R_I = \dfrac{1}{a_1} \times$ $\dfrac{r_m l_3 (r_m a_3 + r_c (l_3 - a_3))}{r_m l_3 (l_2 - a_2) + r_m a_2 a_3 + r_c a_2 (l_3 - a_3)}$ and $R_{II} = \dfrac{1}{(l_1 - a_1)} \times$ $\dfrac{(r_m a_3 + r_c (l_3 - a_3))(r_c a_3 + r_p (l_3 - a_3))}{(r_c a_3 + r_p (l_3 - a_3)) a_2 + (r_m a_3 + r_c (l_3 - a_3))(l_2 - a_2)}$
M	$M_i = \dfrac{a_j a_k M_p}{l_j l_k} + \dfrac{l_i M_p M_c ((l_j - a_j) a_k + (l_k - a_k) a_j)}{l_j l_k (a_i M_c + (l_i - a_i) M_p)}$ $+ \dfrac{(l_j - a_j)(l_k - a_k) l_i M_c M_m}{l_j l_k ((l_i - a_i) M_c + a_i M_m)}$	$M_i = \dfrac{a_j a_k M_m}{l_j l_k} + \dfrac{l_i M_m M_c ((l_j - a_j) a_k + (l_k - a_k) a_j)}{l_j l_k (a_i M_c + (l_i - a_i) M_m)}$ $+ \dfrac{(l_j - a_j)(l_k - a_k) l_i M_c M_p}{l_j l_k ((l_i - a_i) M_c + a_i M_p)}$
μ	$\mu_i = \dfrac{l_i \mu_c \mu_m (l_j - a_j)(l_k - a_k)}{l_j l_k ((l_i - a_i)\mu_c + a_i \mu_m)}$	$\mu_i = \dfrac{a_j a_k \mu_m}{l_j l_k} + \dfrac{l_i \mu_m \mu_c ((l_j - a_j) a_k + (l_k - a_k) a_j)}{l_j l_k (a_i \mu_c + (l_i - a_i)\mu_m)}$

The results of the vertical subdivision of the isolated pore model for fluid inclusions with aspect ratios of 10^{-1}, 10^{-2} and 10^{-3} are presented in Fig. 6.4 in comparison with the results for isolated spheroidal inclusions given by Semjonov (1948) wherein the parameters for calculations are assumed as $\sigma_m = 10^{-4}$ S/·m, $\sigma_p = 10^{-1}$ S/·m; and geometric parameters of the model: $l_1 = l_2 = l_3 = 1$ mm, $a_1 = a_2 = 0...1$ mm, curve I - $a_3 = a_1/1000$, curve II - $a_3 = a_1/100$, curve III - $a_3 = a_1/10$.

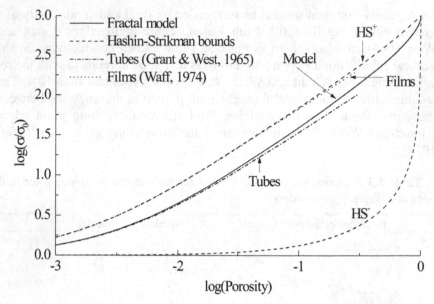

Fig. 6.3 Results of conductivity modeling for the vertical subdivision of Spangenberg model with 0 generation in contact region in comparison with other models

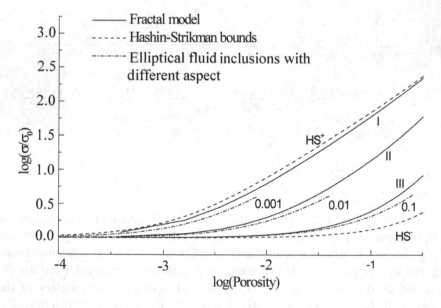

Fig. 6.4 Results of conductivity modeling for the isolated pore model with 0 generation in the contact region in comparison with Semjonov's (1948) results for elliptical fluid inclusions with different aspect ratio

The comparison of electrical conductivities calculated using both open and isolated pore unit demonstrated an agreement of the reviewed model with other theoretical models.

6.5 Application of the fractal model to collocated magnetotelluric and seismic velocity tomography results

The fractal model was applied to the collocated magnetotelluric (MT) and seismic velocity tomography results obtained at Nagamachi-Rifu zone area (Pervukhina et al. 2004).

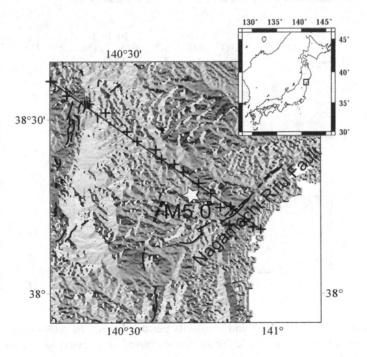

Fig. 6.5 Mutual location of MT sites, seismic velocity tomography results profile and surface trace of the Nagamachi-Rifu fault. The star indicates the epicenter of an earthquake with M 5.0, which occurred on this fault on 15 September 1998 at the depth of about 12 km

A collocated magnetotelluric and seismic velocity tomography experiment was performed to explore a deep extension of the active Nagamachi-Rifu fault, Northern-East Japan. Mutual location of MT points, a line of projection of seismic velocity tomography results and surface trace of Na-

gamachi-Rifu fault are presented in Fig.6.5. The epicenter of an earthquake
with M5.0, which occurred near this fault on 15 September 1998 at the
depth of about 12 km, is indicated by a star (Unimo et al. 2002). They suggested that this event was a slip at a deepest portion of the fault. The seismic velocity tomography images were obtained by Nakajima et al. (2004).

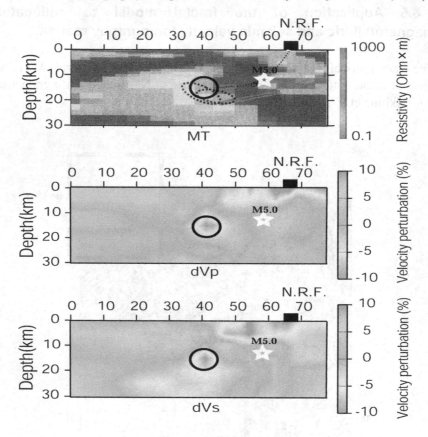

Fig. 6.6 Seismic velocities and resistivity perturbations. The cross-section along
the location of MT sites. Circles show the supposed deep extension of the Nagamachi-Rifu fault

Magnetotelluric image was obtained by Ogawa et al. (2003). Resistivity,
P-wave and S-wave velocity perturbation images are presented in Fig. 6.6.
The parameters of the model for simulation for elastic and electrical properties are presented in Table 6.4 wherein calculations were carried out for
α equals $2 \cdot 10^{-1}$ and 10^{-2} and for t equals $6 \cdot 10^{-4}$, 10^{-2}, $3 \cdot 10^{-2}$, and $9 \cdot 10^{-2}$. Upper subscripts of geometrical sizes indicate generation number. Aspect ra-
tio of pore space of isolated pore generations and the porosity of open pore

generation were chosen as two independent parameters. The calculations were carried out with the next parameters: matrix conductivity $\sigma_m=10^4$ S/m, matrix seismic velocities $V_{p0}=6200$ m/s, $V_{s0}=3570$ m/s, fluid conductivity $\sigma_p=30$ S/m and fluid seismic velocity $V_p=1517$ m/s. For analysis, the grid nodes with compressional velocity V_p perturbations more than 2.5% of the average value in the region of suggested deep extension of Naga-machi-Rifu fault were taken into account. The experimental data of resistivity and seismic velocities were compared with the simulation results for two aspect ratio $\alpha=0.2$ and $\alpha=0.01$ (Fig. 6.7a and b correspondingly). The aspect ratio of 0.2 corresponds to interfacial energy-controlled fluid geometry of porous rock and aspect ratio of 0.01 characterizes regions where pore geometry is far from the dominant textural equilibrium (Takei 2002). For both aspect ratios, calculations were fulfilled for the porosities of the open pore generation equals to 10^{-4} %, $3\cdot10^{-2}$ %, $3\cdot10^{-1}$ % and 2%.

Table 6.4 Geometric parameters of the model for simulation of elastic and electrical properties mentioned in Fig. 6.7

n	Type of generation	l_1 [mm]	l_2 [mm]	l_3 [mm]	a_1 [mm]	a_2 [mm]	a_3 [mm]
0	Isolated pore unit	1	1	1	$l_1^0(1-\alpha i)$, i= 0.01..1/α	$l_2^0(1-\alpha i)$, i= 0.01..1/α	$l_3^0(1-\alpha i)$, if i≤1, 0, if i>1; i= 0.01..1/α
1	Isolated pore unit	$a_1^0/2$	$a_1^0/2$	$a_1^0/2$	$l_1^1(1-\alpha i)$, i= 0.01..1/α	$l_2^1(1-i)$, if i≤1 0, if i>1; i= 0.01..1/α	$l_3^1(1-\alpha i)$, i= 0.01..1/α
2	Isolated pore unit	$a_1^1/2$	$a_1^1/2$	$a_1^1/2$	$l_1^2(1-i)$, if i≤1 0, if i>1; i= 0.01..1/α	$l_2^2(1-\alpha i)$, i= 0.01..1/α	$l_3^2(1-\alpha i)$, i= 0.01..1/α
3	Interconnected pore unit	$a_2^2/2$	$a_2^2/2$	$a_2^2/2$	$l_1^3\cdot t$	$l_2^3\cdot t$	$l_3^3\cdot t$

n is a generation number

An attempt to explain the mutual resistivity and seismic velocities reduction with equilibrium pore geometry (aspect ratio $\alpha=0.2$) (Fig. 6.7a) leads to large values of the porosity of 4-5%. Such values of 4-5% are surprisingly large for the depth of about 15 km.

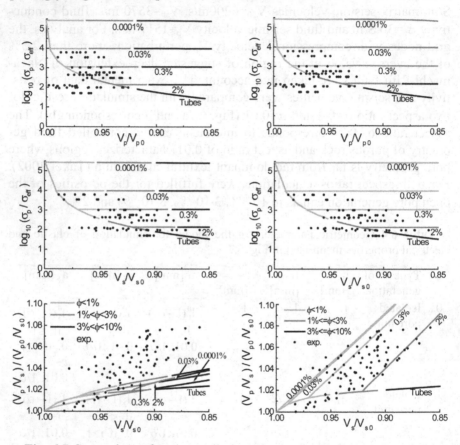

Fig. 6.7 Comparison of magnetotelluric and seismic velocity tomography data (points) with simulation results (lines) for two aspect ratios (a) of 0.2 and (b) of 0.01 (b). Porosity with the open pore generation is indicated along the lines. Different total porosity is presented with different colors: light gray $\phi \leq 1\%$, gray $1\% < \phi \leq 3\%$, and black $\phi > 3\%$

Moreover, the values of the ratio of the measured compressional and shear velocities divided by the ratio of the compressional and shear velocities of the matrix material $(V_p/V_s)/(V_{p0}/V_{s0})$ are larger than the predictable values throughout the range of V_s/V_{s0}, the ratio of the measured shear velocity to the shear velocity of the matrix material (Fig. 6.7a, bottom graph). While, for the small aspect ratio of 0.01, the experimental velocity perturbation can be explained with less than 3% of porosity and measured reduction V_p with V_s is correspond to simulating curves (Fig. 6.7b). As the small aspect ratios of 0.01 means that the pore microstructure is far from equilibrium, this result suggests some strong shear stress regime that preventing the fluid filled pores from achieving the equilibrium geometry. Thus, application of the fractal model for quantitative analysis of experimental data of resistivity and seismic velocities was fruitful for recognition of the state of active fault.

6.6 Discussion

The present review mentioned some deficiency of the mixture theories of elastic and electrical properties of porous rock that are able to describe the variety of probable microstructures of crust and upper mantle. The fractal network models of electrical and elastic properties of porous rock are shown to be more adequate. However, the fractal models of porous rock (Bahr 1997, Spangenberg 1998, Kozlovskaya and Hjeit 2000) allow calculating properties of porous rock for a wide range of microstructures and this can be deficiency in some cases. For instance, Spangenberg (1998) reports, that in the case of isotropy, all the velocity data between the Hashin-Shtrikman upper and lower bounds can be fitted by the use of different self-similar substructure generations. In the case of numerous simulating parameters and shortage of measured ones, it is difficult to discriminate the influence of a particular factor and make reliable conclusion about rock microstructure.

Further progress in this field can be achieved when plausible microstructures for the crust and the upper mantel are defined more exactly, those microstructures are taken into account for developing of theoretical models for both elastic and electrical properties of porous rock, and experimental data are quantitatively interpreted using the theoretical models that reflect those microstructures.

The fractal model suggested by Pervukhina et al. (2003; 2004) has been developed to describe pore microstructures in the lower crust and the upper mantle and deal with parameters (namely, pore aspect ratio and connectivity) that are generally obtained from analyzing seismic velocity tomography and electrical resistivity experimental data. Application of the model to electrical resistivity and seismic velocities data, measured across the deep extension of the Nagamachi-Rifu fault (Northeastern Japan), allowed to make important conclusion about a strong shear stress regime that preventing the fluid-filled pores from achieving the equilibrium geometry.

Acknowledgements

We thank Dr. V.P. Dimri for his invitation to write this review and Dr. Y. Ogawa, Dr. N. Umino and Dr. J.Nakajima for providing us magnetotelluric and seismic velocity tomography data used in this study. This research was partly supported by the "Comprehensive research program on flow and slip process in and below the seismogenic region" sponsored by Ministry of Education, Culture, Sports, Science and Technology (MEXT) of Japan. M. Pervukhina was supported by Japan Science and Technology Agency.

6.7 References

Bahr K (1997) Electrical anisotropy and conductivity distribution functions of fractal random networks of the crust: the scale effect of conductivity. Geophys J Int 130: 649-660

Bayuk IO, Chesnokov EM (1998) Correlation between Elastic and Transport properties of porous cracked anisotropic media. Phys Chem Earth 23: 361-366

Berryman JC (2000) Mixture theories for rock properties. In: Rock Physics & Phase Relations, American Geophysical Union, Washington

Cohen RW, Cody GD, Coutts MD, Abeles B (1973) Optical properties of granular silver and gold films. Phys Rev B8: 3689-3701

Galeener FL (1971) Optical evidence for a network of cracklike voids in amorphous germanium. Phys. Rev Lett 27: 1716-1719

Grant FS, West GF (1965) Interpretation theory in applied Geophysics. McGraw-Hill Book Comp, New York

Gueguen Y David C, Gavrilenko P (1991) Percolation networks and fluid transport in the crust. Geophy Res Lett 18: 931-934

Hashin Z, Shtrikman S (1961) Note on a variational approach to the theory of composite elastic materials. J Franklin Inst 271: 336-341

Hashin Z, Shtricman S (1962) A variational approach to the theory of the effective magnetic permeability of multiphase materials. J Appl Phys 33: 3125-3131

Hill R (1952) The elastic behaviour of crystalline aggregate. Proc Phys Soc London A65: 349-354

Kozlovskaya E, Hjeit SE (2000) Modeling of Elastic and Electrical Properties of Solid-Liquid Rock System with Fractal Microstructure. Phys Chem Earth (A) 25: 195-200

Kuster GT, Toksöz MN (1974) Velosity and attenuation of seismic waves in two-phase media: Part II Experimental results. Geophysics 39: 607-618

Madden TR (1976) Random networks and mixing laws. Geophysics 41: 1104-1125

Madden TR (1983) Microcrack connectivity in rocks: a renormalization group approach to the critical phenomena of conduction and failure in crystalline rocks. J Geophys Res 88: 585-592

Nakajima J, Hasegawa A, Horiuchi S, Yoshimoto K, Yoshida T, Umino N (2004) Crustal heterogeneity around the Nagamachi-Rifu fault, northeastern Japan, as inferred from travel-time tomography. Earth Planets Space (submitted)

Ogawa Y, Mishina M, Honkura Y, Takahashi K, Tank SB (2003) The 1998 M5.0 Main Shock Region in Sendai, Japan, Imaged by magnetotellurics, American Geophysical Union 2003 Fall meeting

Pervukhina M, Kuwahara Y, Ito H (2003) Modelling of elastic and electrical properties of dry and saturated rock. XXIII General assembly of international union of geodesy and geophysics, June 30 – July 11 Sapporo, Japan

Pervukhina M, Ito H, Kuwahara Y (2004) Rock microstructure in the deep extension of the Nagamachi-Rifu fault revealed by analysis of collocated seismic
   and magnetotelluric data: implication of strong deformation process. Earth
   Planets Space (accepted)
Reuss A (1929) Die Berechnung der Fliessgrenze von Mischkristallen auf Grund
   der Plastizitätsbedingungen für Einkristalle. Z Angew Math Mech 9: 49-58
Semjonov AS (1948) An effect of structure of the specific resistance of compos-
   ites. Theme: Resistance of rocks and ores. Vsesoyusnii nauchno-
   issledovatel'skii geologicheskii institut Leningrad. Materiali Seriya Geo-
   phisika 12: 43-61 (in Russian)
Shermergor TD (1977) Theory of elasticity of inhomogeneous media. Moscow:
   Nauka (in Russian)
Spangenberg E (1998) A fractal model for physical properties of porous rock:
   Theoretical formulations and applications. J Geophys Res 103: 12269-12289
Takei Y (2002) Effect of pore geometry on V_p/V_s: From equilibrium geometry to
   crack. J Geophys Res 107: 10.1029/2001JB000522
Umino N, Ujikawa H, Hori S, Hasegawa A (2002) Distinct S-wave reflectors
   (bright spots) detected beneath the Nagamachi-Rifu fault, NE Japan. Earth
   Planets Space 54: 1021-1026
Voigt W (1910) Lehrbuch der Kristallphysik. Teubner-Verlag, Leipzig, Germany
Waff HS (1974) Theoretical considerations of electrical conductivity in a partially
   molten mantle and implications for geothermometry. J Goephys Res 79: 4003-
   4010
Walpole LJ (1969) On the overall elastic moduli of composite materials. J Mech
   Phys Solids 17: 235-251
Walsh JB (1969) New analysis of attenuation in partially melted rock. J Geophys
   Res 74: 4333-4337
Wu TT (1966) The effect of inclusion shape on the elastic moduli of a two-phase
   material. Int J Solids Struct 2: 1-8

Chapter 7. Scaling Evidences of Thermal Properties in Earth's Crust and its Implications

V.P. Dimri, Nimisha Vedanti

National Geophysical Research Institute, Hyderabad, India

7.1 Summary

Fractal behaviour of the Earth's physical properties has been discussed briefly in chapter 1. In this chapter, thermal properties of the Earth's crust are analyzed and the significance of the results obtained is discussed. Here we redefine the traditional heat conduction equation for computation of geotherms by incorporating fractal distribution of thermal conductivity. Further, our study suggests the fractal distribution of radiogenic heat production rate inside the Earth, against the popularly used exponential and step models, which needs to be incorporated in the heat conduction equation.

7.2 Introduction

Geophysical observations made at specific time/space interval form the time series. Two main attributes, governing the properties of data series are statistical distribution of values in the series and the correlation between the values, which is known as persistence. All values in white noise data are independent of other values, hence the persistence becomes zero. The time series that exhibit long-range persistence is termed as fractal time series. Appropriate tools to analyze the variability and the degree of persistence within data sets are power spectra, rescaled range analysis (R/S) analysis and wavelet analysis. In this chapter, we briefly explain the basic principles of all these techniques and apply them to understand thermal structure of the Earth's crust.

7.2.1 Spectral analysis

Spectral analysis is a technique that estimates the power spectral density function, or power spectrum, of a time series. The standard approach is to carry out a Fourier transform on a time series, which is concerned with approximating a function by a sum of sine and cosine terms (Robinson and Trietel 1980, Pristley 1989, Dimri 1992, Percival and Walden 1993). Although the Fourier spectral analysis is the most popular method of spectral estimation, it suffers with major drawbacks like loss of temporal localization, Gibbs phenomena, and elimination of some portion from all frequencies when the signal is transformed back to time domain.

7.2.2 Rescaled range analysis

Hurst et al. (1965), found empirically that many data sets in nature satisfy the power law relation given as

$$\frac{R_K}{S_K} = \left(\frac{K}{2}\right)^{Hu} \tag{7.1}$$

where H_u is the Hurst coefficient, K is the lag for which a range (R_K) and standard deviation (S_K) of the data set are to be calculated.
The range (R_K) is defined as

$$R_K = (y_n)_{max} - (y_n)_{min,} \tag{7.2}$$

where

$$y_n = \sum_{i=1}^{n} (y_i - \bar{y}_n), \; n = 2, N-1 \tag{7.3}$$

wherein y represents values of the time series and n is the number of data points.

The plot of log(R/S) vs. log (K) is a straight line whose slope gives the value of Hurst coefficient H_u. The robustness of the R/S analysis is advocated by Mandelbrot and Wallis (1969), as compared to the spectral methods. The characteristic measure of R/S analysis is the Hurst coefficient. This method quantifies the strength of persistence in the range $0 < H_u < 1$. White noise has Hurst coefficient $H_u = 0.5$. Values of $H_u > 0.5$ reveal persistence of the signal from smaller to larger scale, whereas $H_u < 0.5$ indicates anti-persistence in a given data sequence (Mandelbrot 1982).

7.2.3 Wavelet analysis

In many ways wavelet analyses are the most satisfactory measure of the strength of persistence or fractal behaviour, especially in case of non-stationary time series (Malamud and Turcotte 1999). In wavelet analysis, the data series is convolved with the series of wavelet filters. The generalized form of wavelet transform is given by

$$W(x,y) = \frac{1}{\sqrt{|x|}} \int_{-\infty}^{\infty} f(t)\Psi\left(\frac{t-y}{x}\right)dt \tag{7.4}$$

where $\Psi\left(\dfrac{t-y}{x}\right)$ is a wavelet and $W(x, y)$ are wavelet coefficients generated as a function of x and y, wherein 't' is time, 'x' is time scaling or dilation and, 'y' is time shift or translation. Repeating patterns in wavelet absolute coefficient plot of the time series reveal fractal behaviour in the data.

7.3 Fractal behaviour of thermal conductivity data

The rock thermal conductivity depends upon many parameters like mineralogy, porosity, temperature, pressure, pore fluid of the subsurface rocks etc. An accurate estimation of thermal conductivity is essential to compute geo-isotherms inside the Earth. In this study we use the thermal conductivity data of core samples collected at different depths from the German continental deep drilling project (KTB) borehole as they represent more realistically the insitu conditions. The time series of thermal conductivity with depth contains all information regarding the variation of thermal conductivity with depth, which in turn depends upon many other factors like genesis of crust and pressure-temperature regime. Since it is difficult to measure the variation in thermal conductivity caused by these parameters separately, we study the combined effect of all these parameters by the statistical methods discussed above. These methods are useful in understanding the heterogeneity of the Earth system.

We carried out the R/S and wavelet analysis of the KTB core samples and the result obtained from the R/S analysis is shown in Fig.7.1. The value of Hurst coefficient (0.8 ± 0.048) confirms fractal distribution of thermal conductivity in the Earth's crust. Wavelet absolute coefficients plot of thermal conductivity data from the KTB borehole is shown in Fig. 7.2. Repeating patterns as seen in Fig. 7.2, are clear evidences of fractal distribution of the data. Statistical correlation among the values is the characteristic property of a fractal time series; hence the future values of the time series can be forecasted using the Hurst coefficient. Since the upper crustal samples obtained from the boreholes, solely can't represent the lower crust, we analyse only the statistical correlation among the values of the upper crustal samples.

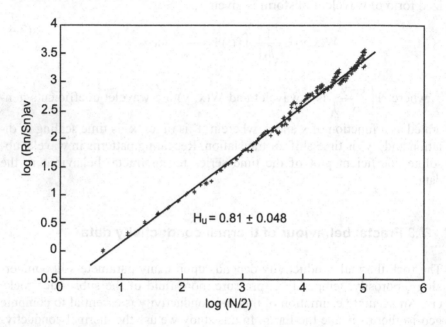

Fig. 7.1 Rescaled range analysis of thermal conductivity data with depth from KTB borehole Germany

To map the fractal thermal conductivity distribution inside the Earth, we assume a model of heterogeneous crust lying over the homogeneous mantle where heterogeneity in crust decreases with depth. An empirical function for estimation of thermal conductivity can be formulated using this model. Here we propose an empirical scaling function to model the effective thermal conductivity in subsurface, in which effective thermal conductivity is given as a product of surface thermal conductivity K and a variable constant 'a' raised to the power 'D', as:

$$K_e = K a^D \qquad (7.5)$$

where 'D' is the fractal dimension of thermal conductivity data, which can be computed as:

$$D = E + 1 - Hu \qquad (7.5a)$$

where in E is Euclidean dimension of the data.

Since, the Earth is more homogeneous at deeper levels; the fractal dimension increases with depth (Nimisha and Dimri 2003). The fractal dimension in Eq. (7.5) accounts for increase in thermal conductivity because of pressure, compaction etc. The variable constant 'a' in Eq. (7.5) accounts for variation in thermal conductivity due to temperature but is assumed to be constant in a particular geological layer. As the effect of temperature on thermal conductivity of basalts is of relatively small magnitude, the value of 'a' for middle and lower crust comes to 1 using Eq. (7.5). Change in the value of fractal dimension and the variable constant 'a' at deeper levels results change in value of K_e at boundaries (Nimisha and Dimri 2004).

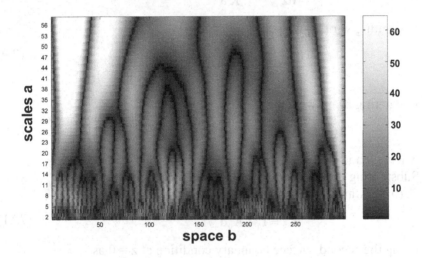

Fig. 7.2 Wavelet absolute coefficient plot of thermal conductivity data with depth from KTB borehole Germany.

7.4 Application to 1-D heat conduction problem

Keeping in view the fractal behaviour of thermal conductivity, we reformulate 1-D heat conduction equation. Following Lachenbruch (1970) and Lachenbruch and Sass (1978), the temperature-depth distributions can be computed as:

$$\frac{d}{dz}(K(z)\frac{dT}{dz}) = -A_0\, e^{-Z/d} \qquad (7.6)$$

where T is temperature as a function of depth z, A_0 is radiogenic heat production, and d is characteristic depth.

By substituting Eq. (7.5) in Eq. (7.6) and considering the variation of thermal conductivity independent of depth in a layer, we get

$$K\, a^D\, \frac{d^2\, T}{dz^2} = -A_0\, e^{-Z/d} \qquad (7.7)$$

or

$$\frac{d^2\, T}{dz^2} = \frac{-A_0\, e^{-Z/d}}{K\, a^D} \qquad (7.8)$$

Integrating w.r.t. z, we get

$$\frac{dT}{dz} = \frac{A_0\, d\, e^{-Z/d}}{K\, a^D} + C_1 \qquad (7.9)$$

Integrating again we get

$$T(z) = -A_0\, d^2\, e^{-Z/d}/K\, a^D + C_1 z + C_2 \qquad (7.10)$$

where C_1 and C_2 are integration constants.

Substituting the first surface boundary condition at z = 0 as T (0) = T_0 and D = 0, in Eq. (7.10) we get

$$C_2 = T_0 + A_0 d^2/K \qquad (7.11)$$

Using the second surface boundary condition at z = 0 as

$$K\, a^D\, \frac{dT}{dz} = Q_S \text{ and } D = 0$$

in Eq.(7.9), we get

$$C_1 = \frac{Q_S - A_0 d}{K} \qquad (7.12)$$

Substituting values of C_1 and C_2 in Eq. (7.10) we get

$$T(z) = T_0 + \frac{Q_S z}{K} + \frac{A_0 d^2}{K}\left(1 - \frac{e^{-Z/d}}{a^D} - \frac{z}{d}\right) \qquad (7.13)$$

The Eq. (7.13) can be used to obtain the crustal temperature depth distributions.

Below Moho Eq. (7.6) becomes zero hence we get

$$K a^D \frac{d^2 T}{dz^2} = 0 \qquad (7.14)$$

Integrating the Eq. (7.14) we get

$$K a^D \frac{dT}{dz} = C_3 \qquad (7.15)$$

where C_3 is integration constant.

Integrating Eq. (7.15) we get

$$T(z) = \frac{C_3 z}{Ka^D} + C_4 \qquad (7.16)$$

where C_4 is integration constant.

Eq.(7.16) at $z = z_m$ and $T(z) = T_m$ yields

$$T_m = \frac{C_3 z_m}{Ka^d} + C_4 \qquad (7.17)$$

where z_m is depth to Moho and T_m is temperature at Moho.

The boundary condition at $z = z_m$ is given as

$$K a^D \frac{dT}{dz} = Q_m \qquad (7.18)$$

where Q_m is mantle heat flow

From Eqs (7.15) and (7.18), we get

$$C_3 = Q_m \qquad (7.19)$$

Substituting (7.19) in (7.15) we get

$$C_4 = T_m - \frac{Q_m z_m}{Ka^D} \qquad (7.20)$$

Again substituting the values of C_3 and C_4 in Eq.(7.16) we get

$$T(z) = \frac{Q_m}{Ka^D}(z - z_m) + T_m \qquad (7.21)$$

The Eq.(7.21) can be used to quantify the lithospheric temperature distributions below Moho.

7.5 Source distribution of radiogenic heat generation inside the Earth

In the continental crust, temperature distribution depends mainly on two factors; the heat supplied from the earth's interior, and the radiogenic heat sources present in the upper crust. The crustal radiogenic sources contribute to about 40% of the heat flow at the surface in continental regions making this an important element in determining the crustal temperature distribution. ^{238}U and ^{232}Th are the primary heat producing elements today. ^{235}U and ^{40}K were more important in the Earth's early history.

Most radioactive isotopes are concentrated in the upper continental crust. There is very small amount of radioactive isotopes in the oceanic crust and almost negligible in the mantle. In the absence of direct measurements of radioactive heat generation, various models for the depth distribution of radiogenic heat sources have been proposed in the literature (Birch et al 1968, Lachenbruch 1970, Lachenbruch and Bunder 1971). Among the various models proposed for the depth distribution of radiogenic sources, the exponential model (Lachenbruch 1970, Lachenbruch and Sass 1978) has been widely used. Heat production in the crustal rocks is caused by the transformation of kinetic energy to thermal energy during the radioactive decay of uranium, thorium, and potassium. In general the upper crust is composed of granites. The range of typical heat production rate for granite or equivalent rocks is given as $2\mu Wm^{-3}$- $10\mu Wm^{-3}$.

The variation of heat production rate in granites with depth has been studied for the boreholes GPK1 and GPK2, situated at the European Hot-Dry-Rock site in Soultz (Fig. 7.3). Lachenbruch (1968; 1970) suggested an exponential decrease of heat production rate in granitic plutons with restriction to regions showing a linear heat flow-heat production relation (Pribnow and Winter 1997). Dashed line in Fig.7.3 indicates an exponential fit to the data. Deviation from exponential model fit is observed at greater depths in Fig. 7.3. If we zoom the lower part of figure, this argument becomes clearer that at greater depths, heat production rate in this borehole cannot be explained by the same exponential model.

We have carried out rescaled range analysis of the calculated heat production data of Soultz boreholes. The value of Hurst coefficient for the data is obtained as 0.84 ± 0.073 (Fig.7.4), which clearly indicates the fractal distribution of the heat production data.

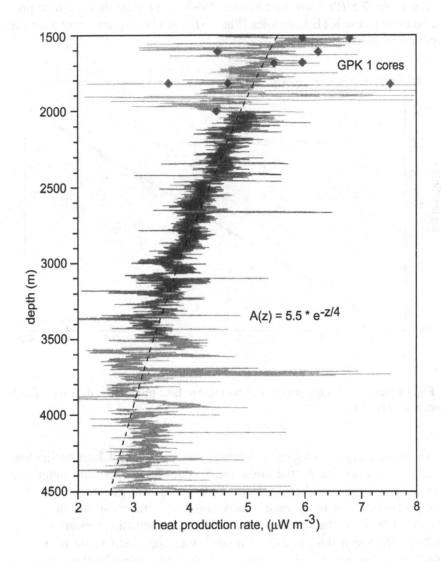

Fig. 7.3 Heat production rate calculated from Natural Gamma-Ray Spectrometer (NGS) logs run by Schlumberger in GPK1 and GPK2 boreholes of Soultz. The dashed line indicates an exponential fit (after Pribnow et al. 1999)

Another evidence of complex behaviour of heat production rate is shown in Fig.7.5 (Pribnow and Winter 1997). It is clear that the heat production rate in the KTB boreholes (Fig.7.5), can't be explained by a simple exponential model.

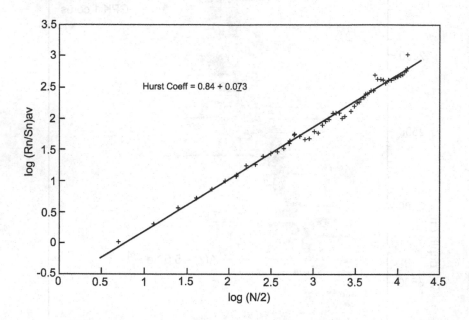

Fig.7.4 Rescaled Range analysis of radiogenic heat production data with depth from Soultz borehole.

An opportunity to study more directly the variation of heat production with depth is provided by the lower crustal data from Western Canterbury region, New Zealand. The variation of measured heat generation with the depth corresponding to the expressed vertical sequence is shown in Fig.7.6 (Pandey 1981). At the first look the heat production rate appears to vary randomly between 0.92 and 2.39 $\mu w/m^3$ with the mean value of 1.56 $\pm$ 0.007 $\mu w/m^3$, which is very much higher than the normal value taken for the lower crust. This contributes an important evidence for the deviation of the exponential model of heat production rate of lower crustal rocks. Rate of decrease in heat production with depth is necessary to consider the variations of heat flow with depth and heat flow extrapolation to greater depth is crucial to obtain the boundary conditions in numerical models. Therefore this study has implications to the temperature distribution in the crust and the results thus obtained indicate a radiogenic source distribution which is more complex than a simple exponential model. This finding also needs to be incorporated in the heat conduction equation (Eq. 7.5) to compute the crustal geotherms.

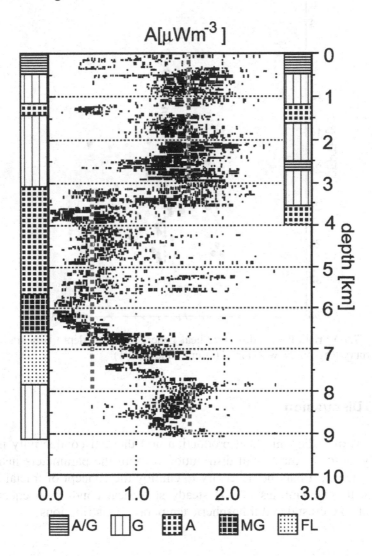

Fig. 7.5 Heat production data from borehole and laboratory measurements for KTB borehole with simplified lithological profiles for the HB (left) and VB (right) A: Amphibolite; MG: Metagrabro G: Gneiss A/G: Alternate sequence FL: Franconian Lineament. Dashed line represents the average values for respective lithology (after Pribnow and Winter 1997)

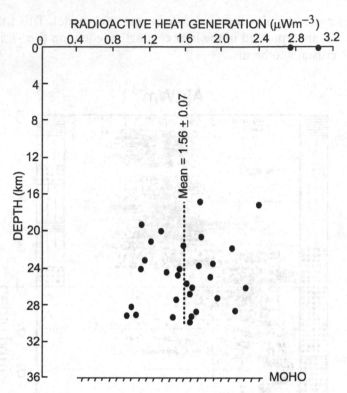

Fig. 7.6 Variation of radioactive heat generation with depth beneath Western Canterbury Region, New Zealand (after Pandey 1981)

7.6 Discussion

Analysis of radiogenic heat production and thermal conductivity in crust clearly indicates the fractal distribution of both the parameters inside the Earth. Hence, it may be necessary to employ the concept of fractal behaviour of these quantities in 1-D steady state heat conduction equation to compute the crustal and lithospheric temperature distributions.

Acknowledgements

We acknowledge authors of GGA technical report on Heat production and temperature to 5 km depth at the HDR site in Soultz-sous-Forets, authors and American Geophysical Union publishers for reproduction of figures 7.3 and 7.5 respectively.

7.7 References

Birch F, Roy RF, Decker ER (1968) Heat flow and thermal history in New York and New England. In: E-an Zen, White WS, Hadley JB, Thompson JB Jr (eds) Studies of Appalachian geology: Northern and maritime, New York, Interscience, pp 437-451

Dimri VP (1992) Deconvolution and inverse theory. Elsevier Science Publishers, Amsterdam London New York Tokyo

Hurst HE, Black RP, Simaika YM (1965) Long-Term Storage: An Experimental Study. Constable, London

Lachenbruch AH (1968) Preliminary geothermal model of the Seirra Nevada. J Geophy Res 73: 6977-6989

Lachenbruch AH (1970) Crustal temperature and heat production: implication of the linear-flow relation. J Geophy Res 75: 3291-3300

Lachenbruch AH, Bunder CM (1971) Vertical gradient of heat production in the continental crust, some estimates from borehole data. J Geophys Res 76: 3852-3860

Lachenbruch A H, Sass J H (1978) Models of an extending lithosphere and heat flow in the basin and range province. Geol Soc Am 152: 209-250

Malamud BD, Turcotte DL (1999) In: Dmowska R, Saltzman B (eds) Advances in geophysics: long range persistence in geophysical time series, Academic Press, San Diego, pp 1-87

Mandelbrot BB (1982) The Fractal Geometry of Nature. WH Freeman and Co, New York

Mandelbrot BB, Wallis JR (1969) Robustness of the Rescaled range R/S in measurement of noncyclic long run statistical dependence. Water Resour Res 5: 967-988

Nimisha V, Dimri VP (2003) Fractal behavior of electrical properties in oceanic and continental crust. Ind Jour Mar Sci 32: 273-278

Nimisha V, Dimri VP (2004) Concept of fractal thermal conductivity in lithospheric temperature studies. Phys Earth Plant Int (submitted)

Pandey OP, (1981) Terrestrial heat flow in New Zealand. PhD Thesis, Victoria Univ. Wellington New Zealand, pp 194

Percival DB, Walden AT (1993) Spectral analysis for physical applications, multitaper and conventional univariate techniques. Cambridge University Press, UK

Pribnow DFC, Winter HR (1997) Radiogenic heat production in the upper third of continental crust from KTB. Geophys Res Lett 24:349-352

Pribnow D, Fesche W, Hagedorn F (1999) Heat production and temperature to 5km depth at the HDR site in Soultz-sous-Forets. Technical report GGA, Germany

Pristley MB (1989) Spectral analysis and time series. Academic Press, London

Robinson EA, Treitel S (1980) Geophysical signal analysis. Prentice Hall Inc, NJ

Chapter 8. Fractal Methods in Self-Potential Signals Measured in Seismic Areas

Luciano Telesca, Vincenzo Lapenna

Institute of Methodologies for Environmental Analysis, Tito (PZ), Italy

8.1 Summary

The self-potential (SP) signals are mainly generated by the streaming potential, which causes a voltage difference when a fluid flows in a porous rock. In seismic focal areas, this effect is strengthened by the increasing accumulation strain, which can produce dilatancy of rocks. Therefore, tectonic processes can be directly revealed by the investigation of the temporal fluctuations of SP signals, which may be useful to monitor and understand complex phenomena related with earthquakes.

Can the concept of fractal be used to qualitatively and quantitatively characterize an SP signal? Fractals are featured by power-law statistics, and, if applied to time series, can be a powerful tool to investigate their temporal fluctuations, in terms of correlations structures and memory phenomena. In the present review we describe monofractal and multifractal methods applied to SP signals measured in seismic areas. Persistent scaling behaviour characterizes SP signals, which, therefore, are not realizations of a white noise process. Furthermore, in multifractal domain SP signals measured in intense-seismicity areas and those recorded in low-seismicity areas are discriminated.

8.2 Introduction

Seismic research has given growing evidence to the analysis of a large variety of geophysical signals that can provide indirect information on the dynamics underlying tectonic processes. Geophysical parameters may be useful to monitor and understand many seemingly complex phenomena linked to seismic activity (Rikitake 1988, Zhao and Qian 1994, Park 1997, Martinelli and Albarello 1997, Di Bello et al. 1998, Vallianatos and Tzanis 1999, Hayakawa et al. 2000, Telesca et al. 2001, Tramutoli et al. 2001). Variations in the stress and fluid flow fields can produce changes in the self-potential (SP) field (Scholz 1990), so that investigating these induced fluctuations may give information on the governing mechanisms both in normal conditions and during intense seismic activity.

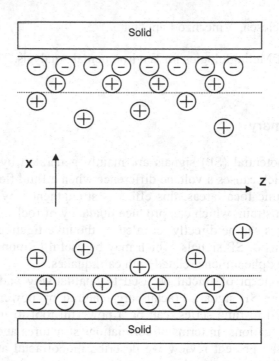

Fig. 8.1 Schematic representation of the double electrical layer

The method of SP consists in measuring the voltage difference between two points on earth's surface due to the presence of an electric field produced by natural sources distributed in the subsoil (e.g. Parasnis 1986, Sharma 1997 and references therein).

The SP signals have been largely applied in geothermal, environmental and engineering research to locate and delineate sources associated with movement of fluids and groundwater (Ogilvy et al. 1969, Corwin and Hoover 1979, Sharma 1997, and references therein). Furthermore, other significant applications can be found in the geophysical survey of volcanic and tectonic areas (Di Maio and Patella 1991, Lapenna et al. 2000, Di Maio et al. 1997).

In near surface geophysics the most relevant phenomenon that could originate the SP anomalous field is known as electrofiltration or streaming potential: the electrical signal is produced resulting from fluid flow in a porous rock under a pore pressure gradient. The phenomenon is generated by the formation within the porous ducts of a double electrical layer between the bounds of the solid that absorbs electrolytic anions and cations distributed in a diffused layer near the boards. The dissolved salts increase the amount of anions and cations of the underground liquids. The free liquid in the centre of the rock pore is usually enriched in cations, while anions are usually absorbed on the soil surface in silicate rock. The free pore water carries an excess positive charge, a part of which accumulates close to the solid-liquid interface forming a stable double layer. When the liquid is forced through the porous medium, the water molecules carry free positive ions in the diffusion part of the pore (Fig. 8.1). This relative movement of cations with reference to the firmly attached anions generates the well known streaming potential (Keller and Frischknecht 1966). Of course, the role of the electrical charges can be reversed, according to the absorption properties of the rocks. As suggested by Mizutani et al. (1976), the streaming potential can be responsible for the voltage measures on the ground surface preceding an earthquake (Patella 1997). Other possible factors are temperature gradients, especially in volcanic areas, and concentration gradients related to tortuousity and narrowings of the capillary system of cracks (Di Maio and Patella 1991).

The increasing accumulation of strain in a seismic focal region can cause dilatancy of rocks (Nur 1972). The phenomenon of dilatancy consists in the formation and propagation of cracks inside a rock as stress reaches about half its strength (Brace et al. 1966). If the rocks volumes in the focal region and surroundings are saturated with fluids, the voids generate pressure gradients, to which the fluid particles are subjected. Hence, fluids invade the newly opened voids and flow until the pressure balances inside the whole system of interconnected pores. During fluid invasion the condition of rock dilatancy hardening can be reached: the rock suddenly weakens and the earthquake is triggered.

In order to assess the use of SP signals as indicators of earthquake preparation (Hayakawa and Fujinawa 1994, Hayakawa 1999), the fundamental issue to address is if these parameters are able to reveal dynamical characteristics of active tectonics, and we have to understand if there is a significant correlation between seismic sequences and SP temporal fluctuations. Obviously, the existence of such a correlation can be established only after a dynamical characterization of these signals has been performed.

To quantitatively characterize SP dynamics, we employ techniques, which are able to extract robust features hidden in their complex fluctuations. Fractality is one of the features of such complexity. What does fractality mean? A fractal is an object whose sample path included within some radius scales with the size of the radius. It is clear from the definition of fractal, that fractal processes are characterized by scaling behaviour, which leads naturally to power-law statistics. In fact, consider a statistics $f(x)$, which depends continuously on the scale x, over which the measurements are taken. Suppose that changing the scale x by a factor a, will effectively scale the statistics $f(x)$ by another factor $g(a)$, $f(ax)=g(a)f(x)$. The only nontrivial solution for this scaling equation is given by $f(x)=bg(x)$, $g(x)=x^c$, for some constants b and c (Thurner et al. 1997, and references therein). Therefore, power-law statistics and fractals are very closely related concepts.

The fractality of a signal can be investigated aiming to its geometrical characterization as self-similar curve; but if the fractality of a time series is studied in order to characterize its temporal fluctuations, we need to perform second-order fractal measures, which furnish information regarding the correlation properties of a time series.

8.3 Power spectrum analysis

The spectral analysis represents the standard method to detect correlation features in time series fluctuations. The power spectrum is obtained by means of the Fourier transform of the signal. It describes how the power is concentrated at various frequency bands. Thus, the power spectrum reveals periodic, multiperiodic or non-periodic signals. The fractality of a time series is revealed by a power-law dependence of the spectrum upon the frequency, $S(f) \sim 1/f^{\beta}$, where the scaling (spectral) exponent β informs on the type and the strength of the time-correlation structures intrinsic in the signal fluctuations (Havlin et al. 1999). If $\beta=0$ the temporal fluctuations are purely random, typical of white noise processes, characterized by completely uncorrelated samples. If $\beta>0$, the temporal fluctuations are persistent, meaning that positive (negative) variations of the signal will be very likely followed by positive (negative) variations; this feature is typical of system which are governed by positive feedback mechanisms. If $\beta<0$, the temporal fluctuations are anti-persistent, meaning that positive (negative) variations of the signal will be very likely followed by negative (positive) variations; this feature is typical of system which are governed by negative feedback mechanisms.

Generally, observational time series are affected by gaps; thus, they can be considered as unevenly sampled. Therefore, for unevenly sampled time-series the power spectrum can be calculated by means of the Lomb periodogram method (Lomb 1976).

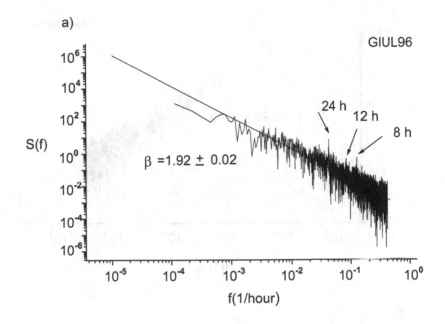

a)

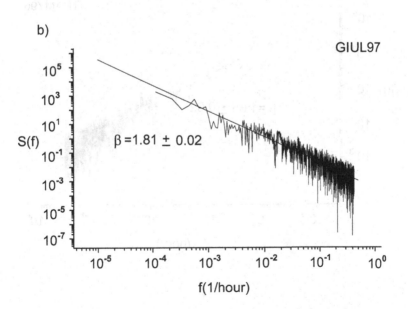

b)

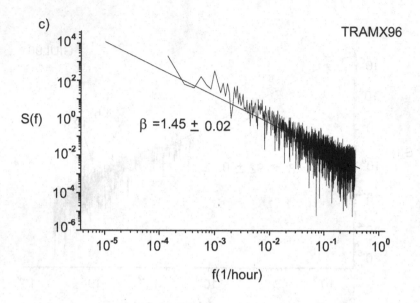

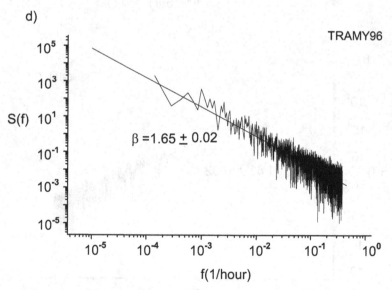

Fig. 8.2 (a-d) Lomb spectra of four SP signals measured in southern Italy. The spectra reveal that the signals are not characterized by white noise dynamics, but present long-range correlations, as indicated by the scaling behaviour (after Colangelo et al. 2003)

Denoting as x_n the datum measured at instant t_n, the Lomb Periodogram is defined by the following formula:

$$P(\omega) = \frac{1}{2\sigma^2} \left\{ \frac{\left[\sum_n (x_n - \bar{x})\sin\omega(t_n - \tau)\right]^2}{\sum_n \sin^2\omega(t_n - \tau)} + \frac{\left[\sum_n (x_n - \bar{x})\cos\omega(t_n - \tau)\right]^2}{\sum_n \cos^2\omega(t_n - \tau)} \right\} \qquad (8.1)$$

where $\omega = 2\pi f$ is the angular frequency and τ is given by

$$\tan(2\omega\tau) = \frac{\sum_n \sin 2\omega t_n}{\sum_n \cos 2\omega t_n}. \qquad (8.2)$$

The slope of the line fitting the log-log plot of the power spectrum by a least square method in the linear frequency range gives the estimate of the spectral index β. Fig.8.2 provides an example in which the Lomb Periodogram of four SP is calculated. The power-law behaviour extending over a wide range of frequencies in all signals denotes that they are long-range correlated and not realizations of white noise processes, which model uncorrelated systems. The power spectrum, furthermore, evidences the presence of some periodic components superimposed on the basic power-law behaviour, thus informing on the coexistence of different mechanisms (meteo-climatic-type), which modulate the SP variability at particular frequencies (Fig. 8.2a). Dimri and Ravi Prakash (2001) have used the Lomb approach (1976) to get power spectrum of unevenly spaced fossil records.

The estimate of the spectral exponent is rather rough, due to large fluctuations in the power spectrum, especially at high frequencies. Furthermore, the power spectrum is sensitive to nonstationarities that could be present in observational data.

8.4 Higuchi analysis

In the literature, many papers have been devoted to find methods capable of giving stable estimations of the power-law spectral index. Burlaga and Klein(1986), presented a method to calculate stable values of the fractal dimension D of large-scale fluctuations of the interplanetary magnetic field; the relationship between the fractal dimension D and the spectral exponent β, is given by Berry's expression $D = (5-\beta)/2$ (Berry 1979), for $1 < \beta < 3$. They defined the length $L_{BK}(\tau)$ of the B(t) curve as

$$L_{BK}(\tau) = \sum_{k=1}^{N} \left| \overline{B}(t_k + \tau) - \overline{B}(t_k) \right| / \tau \tag{8.3}$$

where $\overline{B}(t_k)$ denotes the average value of B(t) between $t=t_k$ and $t=t_k+\tau$. This length is a function of τ, and for statistically self-affine curves, the length is expressed as $L_{BK}(\tau) \propto \tau^{-D}$. Using this relation, the value of D can be estimated as the slope of the log-log plot of the length $L_{BK}(\tau)$ versus the time interval τ. Then, using Berry's relation $D=(5-\beta)/2$ for $1<\beta<3$, the spectral exponent can be estimated.

Another method, which also gives a stable value of the fractal dimension, has been presented by Higuchi (1988; 1990). A new time-series is constructed from the given time series X(i), (i=1, 2, ..., N),

$$X_\tau^m; X(m), X(m+\tau), X(m+2\tau),...., X(m+[(N-m)/\tau]\tau); (m=1,.....\tau) \tag{8.4}$$

where [ ] implies Gaussian notation. The length of the curve is defined as

$$L_m(t) = \left\{ \left(\sum_{i=1}^{[(N-m)/\tau]} \left| X(m+i\tau) - X(m+(i-1)\tau) \right| \right) \frac{N-1}{[(N-m)/\tau]\tau} \right\} \frac{1}{\tau} \tag{8.5}$$

The average value $<L(\tau)>$ over τ sets of $L_m(\tau)$ is defined as the length of the curve for the time interval τ. If $<L(\tau)> \propto \tau^{-D}$, within the range $\tau_{min} \leq \tau \leq \tau_{max}$ then the curve is fractal with dimension D in this range. He examined the relationship between the fractal dimension D and the power law index β, by calculating the fractal dimension of the simulated time series, which follows a single power-law spectrum density. Even in this case, the spectral exponent estimation could be carried out using Berry's relation.

For $0<\beta<1$, Higuchi (1990), proposed an integrated series of the original time series which is given by:

$$X_\Sigma(j) = \sum_{i=1}^{j} X(i) \qquad (j=1,2,...., N) \tag{8.6}$$

which is an increment process of X(i). The fractal dimension of $X_\Sigma(j)$ can be designated by D_Σ, and the power law index can be denoted by β_Σ. As β_Σ is related to β by $\beta_\Sigma=\beta+2$, D_Σ is expressed as $D_\Sigma=(3-\beta)/2$. Thus, the fractal dimension D_Σ is estimated applying Eq. (8.6) for the length of the integrated time series $X_\Sigma(j)$.

Fig. 8.3 shows the spectral analysis performed using the maximum entropy method (Box and Jenkins 1976), for two SP signals measured in southern Italy.

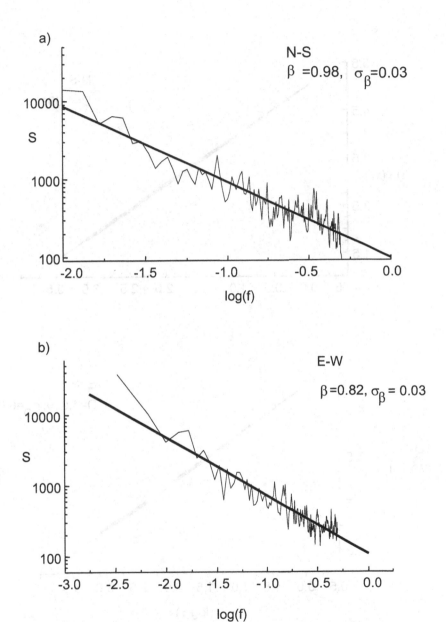

Fig. 8.3 Power spectrum density S(f) vs. frequency f, obtained with maximum entropy method. The two graphs a) and b) are related to two SP signals measured in southern Italy (after Cuomo et al 1999)

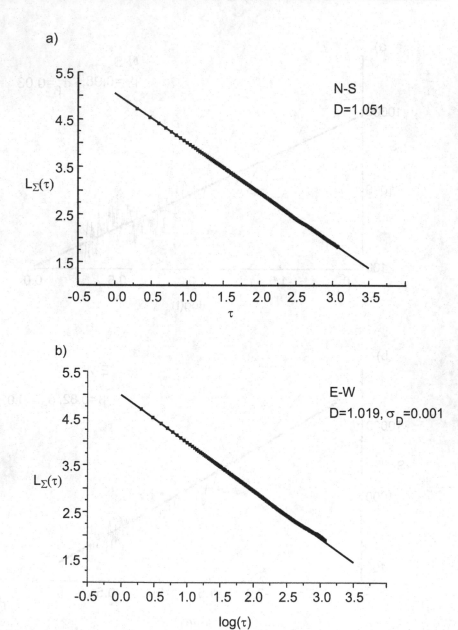

Fig. 8.4 Length of the time series curve L(τ) vs. the time interval τ. The two graphs a) and b) are related to two SP signals measured in southern Italy, whose power spectra is plotted in Figure 8.3 (after Cuomo et al. 1999)

We can exclude the possibility that SP data are purely random (white noise): the analysis shows the presence of the power-law form for the power spectrum density (coloured-noise type), and the slope of the line fitting the log-log plot of the power spectrum density gives an estimate of the spectral index. In both cases, the power spectrum density shows fluctuations that influence the estimate of the power-law index. But, Fig. 8.4 provides an example of the use of the Higuchi method to calculate the fractal dimension D and, thus, the spectral exponent β for the two SP signals in a more accurate manner. The slope of the line fitting the log-log plot of the length of the curve $X_\Sigma(t)$ vs. the time interval τ gives an estimate of the fractal dimension D_Σ, from which the power-law exponents can be calculated.

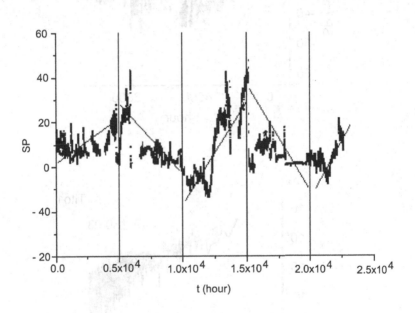

Fig. 8.5 Detrending method performed in the DFA

8.5 Detrended fluctuation analysis

Recently, the method of Detrended Fluctuation Analysis (DFA) has been developed to reveal long-range correlation structures in observational time series. This method was proposed by Peng et al. (1995), and it avoids spurious detection of correlations that are artefacts of nonstationarity, that often affects experimental data. Such trends have to be well distinguished from the intrinsic fluctuations of the system in order to find the correct scaling behaviour of the fluctuations. Very often we do not know the reasons for underlying trends in collected data and we do not know the scales of underlying trends. The DFA is a well established method for determining the scaling behaviour of data in the presence of possible trends without knowing their origin and shape (Kantelhardt et al. 2001).

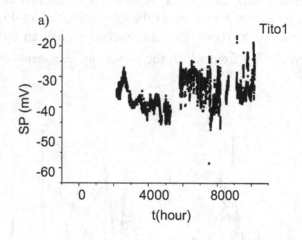

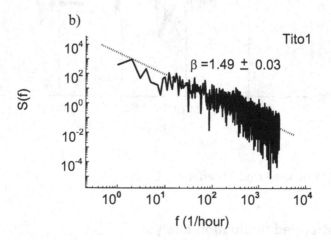

c)

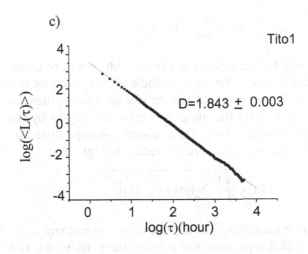

d)

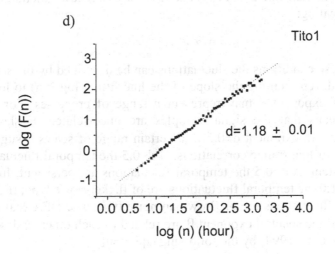

Fig. 8.6 Scaling behaviour in a SP time series (a) investigated by the Lomb Periodogram (b), Higuchi method (c) and DFA (d) (after Telesca et al. 2003)

The methodology operates on the time series x(i), where i=1,2,...,N and N is the length of the series. x_{ave} indicates the average value

$$x_{ave} = \frac{1}{N} \sum_{k=1}^{N} x(k).$$ (8.7)

The signal is first integrated

$$y(k) = \sum_{i=1}^{k} [x(i) - x_{ave}].$$
(8.8)

Next, the integrated time series is divided into boxes of equal length n. In each box a least-squares line is fit to the data, representing the trend in that box (Fig. 8.5). The y coordinate of the straight line segments is denoted by $y_n(k)$. Next we detrend the integrated time series $y(k)$ by subtracting the local trend $y_n(k)$ in each box. The root-mean-square fluctuation of this integrated and detrended time series is calculated by

$$F(n) = \sqrt{\frac{1}{N} \sum_{k=1}^{N} [y(k) - y_n(k)]^2}.$$
(8.9)

Repeating this calculation over all box sizes n, we obtain a relationship between F(n), which represents the average fluctuation as a function of box size, and the box size n. If F(n) behaves as a power-law function of n, data present scaling:

$$F(n) \propto n^d.$$
(8.10)

Under these conditions the fluctuations can be described by the scaling coefficient d, representing the slope of the line fitting log F(n) to log n. The values of exponent d may represent a range of processes. For example d=0.5 means that the signal samples are uncorrelated or short-range-correlated. An exponent d≠0.5 in a certain range of scales n suggests the existence of long-range correlations. If d<0.5 the temporal fluctuations are antipersistent. If d>0.5 the temporal fluctuations are persistent. In particular, if d=1.0 the temporal fluctuations are of flicker-noise type; if d=1.5 the temporal fluctuations are of Brownian noise type. The DFA scaling exponent d and the spectral exponent β are related to each other as described in (Buldyrev et al. 1994) by the following equation:

$$d = \frac{1 + \beta}{2}.$$
(8.11)

Fig. 8.6 provides an example of Lomb Periodogram, Higuchi analysis and DFA concomitantly performed on a SP time series. Power-law behaviours are clearly detected in all the statistics. The Lomb Periodogram shows scaling form, but the estimate of the scaling exponent is strongly affected by the fluctuations especially at high frequency bands (Fig. 8.6b). The estimate of the scaling exponent by means of the Higuchi method presents higher stability. The DFA method furnishes a measure for the scaling exponent slightly different from that given by the spectral method. The reason for this difference is due to the capability of the DFA to capture dynamics in nonstationary signals, while the power spectrum is sensitive to trends and nonstationarities, which could affect the scaling behaviour of a time series. Furthermore, Stanley et al. (1996), argue that there is a theoretical support for the hypothesis that DFA produces less noisy plots. They link DFA to a double summation of the series autocorrelation function, while the power spectrum is derived from a Fourier transformation of the autocorrelation. The double summation acts effectively as a noise filter, providing more accurate estimates.

8.6 Magnitude and sign decomposition method

Recently (Ashkenazy et al. 2001), it was shown that the fluctuations in a dynamical signal could be characterized by two components- magnitude (absolute value) and sign (direction). These two quantities reflect the underlying interactions in a system, and the resulting force of these fluctuations at each moment determines the magnitude and direction of the fluctuations (Ashkenazy et al. 2003).

In this section the two-point correlation (scaling) properties of the magnitude and sign series with long-range correlations are studied. Any power-law long-range correlated time series can be decomposed into two sub-series (Ashkenazy et al. 2001), given the series $x(i)$, define the increments as

$$\Delta = x(i+1) - x(i), \qquad (8.12)$$

then the magnitude sub-series is the absolute value of the increments

$$m(i) = |\Delta x(i)|, \qquad (8.13)$$

while the sign sub-series is the sign of the increments

$$s(i) = sgn(\Delta x(i)), \qquad (8.14)$$

therefore

$$\Delta = sgn(\Delta x(i)) |\Delta x(i)|, \qquad (8.15)$$

Then, to identify the presence of correlations and their type in the sign and magnitude sub-series, the detrended fluctuation analysis (DFA) is performed (see section 8.5). Therefore, the correlation analysis of the magnitude and sign sub-series consists of the following steps:

    i) From the time series $x(i)$ the increments $\Delta x(i)$ are derived.

ii) The magnitude m(i) and sign s(i) sub-series are formed from the increments. If $\Delta x(i)=0$, s(i) can be defined as +1, 0 or –1.

iii) To avoid artificial trends from the magnitude and sign series their respective average values are subtracted.

iv) Since the DFA does not permit accurate estimation of scaling exponents of strong anticorrelated signals (d close to zero), the sub-series are firstly integrated.

v) The DFA is performed on the integrated sub-series. Then, the value obtained of the scaling exponent is $d'=d+1$, where d is the scaling exponent of the original sub-series.

It has been shown that scaling in the magnitude series is related to the non-linear properties of the signal, while scaling in the sign series is related to its linear properties (Ashkenazy et al. 2003). As an example, Fig. 8.7 shows: (a) the original series x(i); (b) the increment series $\Delta x(i)$; (c) the magnitude sub-series m(i), and (d) the sign sub-series s(i) of a time series.

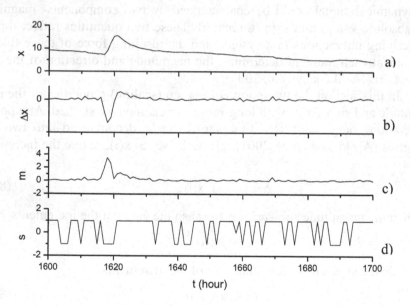

Fig. 8.7 (a) Original series x(i); (b) increment series $\Delta x(i)$; (c) magnitude sub-series m(i), and (d) sign sub-series s(i) of a time series.

Correlation in the magnitude series indicates that an increment with large (small) magnitude is more likely to be followed by an increment with large (small) magnitude. Anticorrelation in the sign series indicates that a positive increment is more likely to be followed by a negative increment and vice versa.

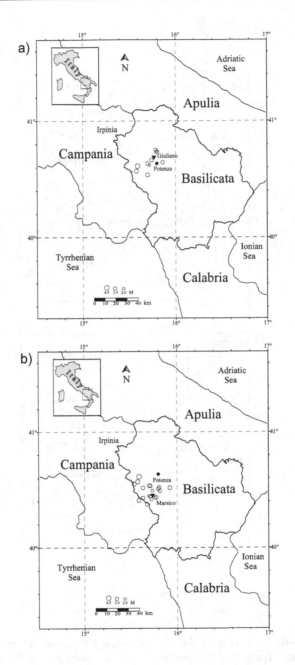

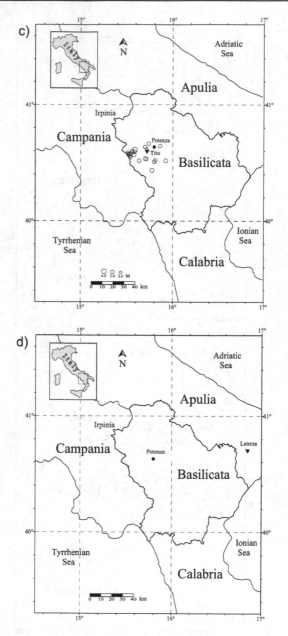

Fig. 8.8(a-d) Ubication of the SP monitoring stations and epicentres of the earthquakes satisfying the Dobrovolsky's rule in relation with the location of the measuring stations: a)Giuliano, b)Marsico, c)Tito and d) Laterza.

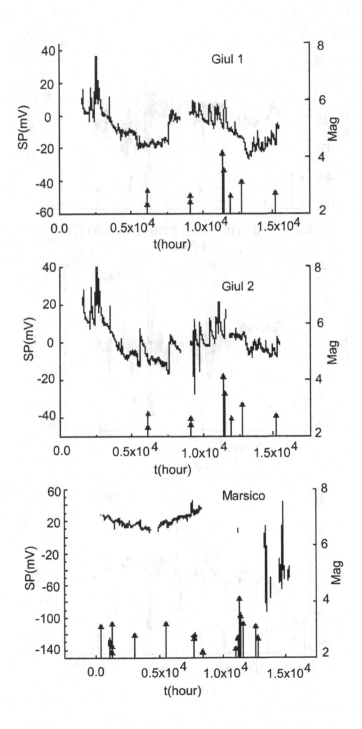

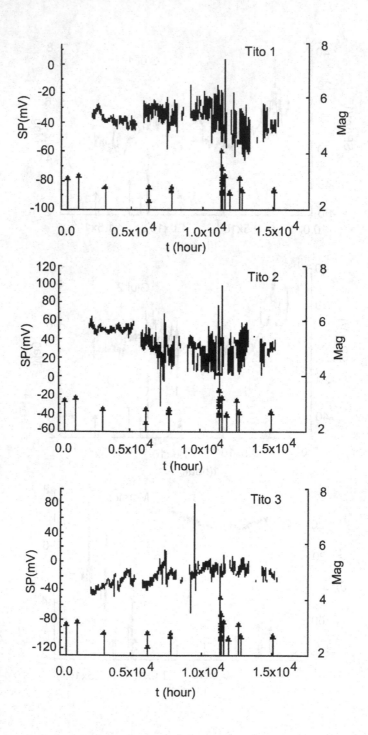

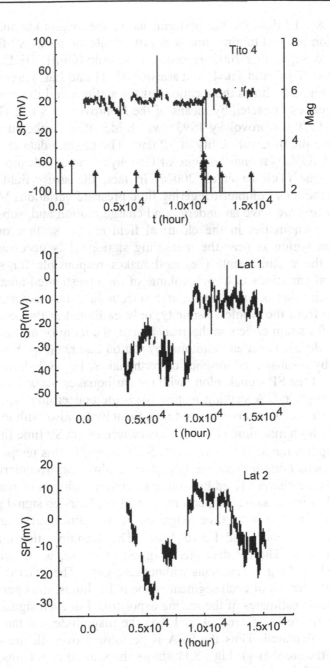

Fig. 8.9 Hourly variability of the nine SP signals recorded at stations Giuliano, Marsico, Tito and Laterza, along with the occurrence of the earthquakes selected by means of the Dobrovolsky's rule (after Telesca et al. 2004a)

Figs. 8.8-8.12 describe the performance of the magnitude and sign decomposition method to nine time series of minute values of SP from January 2001 to September 2002, recorded in seismic (Giul1, Giul2, Marsico, Tito1, Tito2, Tito3 and Tito4) and aseismic (Lat1 and Lat2) areas in southern Italy. Fig. 8.8 shows the locations of the stations and the epicentres of the earthquakes extracted by means of the Dobrovolsky's rule (Dobrovolsky et al. 1979, Dobrovolsky 1993), which identifies earthquakes capable to influence the time variability of SP data. The seismic data are extracted from the INGV (National Institute of Geophysics and Volcanology) seismic catalogue (Telesca et al. 2004a) In fact; the stress field produces cracks on the rock volumes triggering fluid pressure variations. As a result of this process we have an underground charge motion and, subsequently, we observe anomalies in the electrical field on the surface only if the preparation region is near the measuring station. It is necessary to discriminate the useful events (i.e. earthquakes responsible for significant geophysical variations in a rock volume of the investigated area) from all the seismicity that occurred in the area surrounding the measuring station. Therefore, from the whole seismicity, only earthquakes that could be responsible for strain effects in the areas around the monitoring stations have to be considered. The area monitored by station Laterza (Fig. 8.8d) is characterized by a substantial absence of earthquakes. Fig. 8.9 shows the time variations of the SP signals along with the earthquakes occurred during the observation period. A striking feature is visible concerning especially the graphs of station Tito: an increased seismic activity, also with events with a relatively high magnitude ($M \geq 4.0$), characterizes the SP time fluctuations in the temporal range 10^4 hours $< t < 1.5 \cdot 10^4$ hours. In this temporal range, Tito SP fluctuations present the largest variability and irregularity. Giuliano signals are characterized by similar behaviour, with an increased number of spikes in the same temporal range as Tito. Marsico signal presents a long gap during the same time range, but its dynamics vary significantly between $1.3 \cdot 10^4$ hours and $1.5 \cdot 10^4$ hours. The decomposition method has been performed. The SP data present gaps; therefore, we considered for each signal the longest segments without data gaps. The order of the magnitude of the length of each segment is about 10^3 hours, thus permitting to obtain reliable estimates of the scaling exponents. For each signal segment, the increment series is created, and then the magnitude and the sign sub-series are calculated. Thus the DFA is performed over all the sub-series following the steps i)-v). Fig. 8.11 shows the scaling exponents, d_{mag} and d_{sig}, estimated for the magnitude and sign sub-series respectively, for each signal segment.

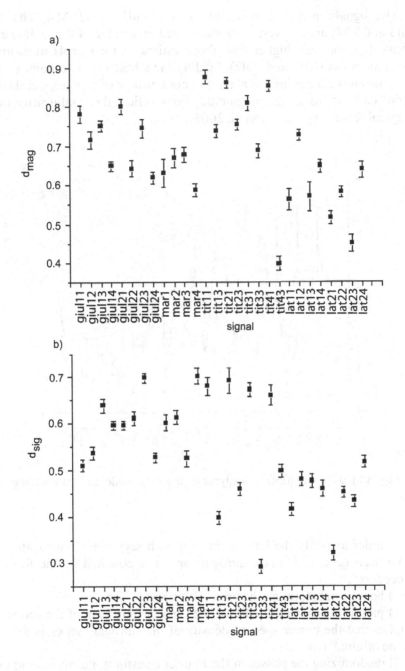

Fig. 8.10 Scaling exponents of the magnitude (a) and sign (b) sub-series for each signal segment

The signals measured in seismic areas (Giul1, Giul2, Mar, Tit1, Tit2, Tit3 and Tit4) are on average characterized by a value of the scaling exponents d_{mag} and d_{sig} higher than those estimated for signals measured in aseismic areas (Lat1 and Lat2). Scaling laws based on two-point correlation methods cannot inform about the nonlinearity of a series, but the two-point correlations in the magnitude series reflect the nonlinearity of the original series (Ashkenazy et al. 2003).

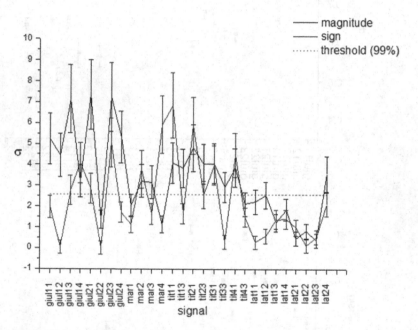

Fig. 8.11 99% significance analysis of the magnitude and sign scaling exponents

In order to verify the latter point, for each segment 10 surrogate series have been generated. Each surrogate series is obtained by the following procedure:

1) Fourier transforming the original series;

2) preserving the amplitudes of the Fourier transform of the series (this implies that the power spectral density of the surrogate series is the same as the original one);

3) randomizing the phases of the Fourier transform, i.e. attributing to the phase a random number between 0 and 2π;

4) inverse-Fourier transforming. The surrogate series, generated by means of this procedure, have the same linear properties as the original ones, like the power spectrum, while the nonlinear properties, stored in the Fourier phases, are destroyed. The difference between the exponents before and after the surrogate data test for nonlinearity may be quantified as follows. If $d_{mag,sig}$ is the exponent derived from the original magnitude or sign sub-series and μ_S, σ_S are the average and standard deviation of the exponents derived from the surrogate data, then the separation is given by

$$\sigma = \frac{\left| d_{mag,sig} - \mu_S \right|}{\sigma_S} \qquad (8.16)$$

where σ measures how many standard deviations the original exponent is separated from the surrogate data exponent. The larger the σ the larger the separation between the exponents derived from the surrogate data and the exponent derived from the original data. Thus, larger σ values indicate stronger nonlinearity. The p-value is calculated by means of the formula $p=erfc(\sigma/\sqrt{2})$ (Theiler et al. 1992), this is the probability of observing σ or larger if the null hypothesis is true, where the null hypothesis is given by the absence of nonlinearity. Fig. 8.11 shows the σ value for the magnitude and sign sub-series for each signal segment. It is also shown by the horizontal dotted line, representing the threshold of 99% significance. Thus, values below this threshold indicate no statistically significant difference between original data and surrogate data; in other word, if the σ value is over the threshold, the corresponding signal has the probability of 99% to be characterized by nonlinear dynamics. We observe that Laterza σ values are below the threshold regarding both the sign and the magnitude series (only Lat24 has the magnitude exponent significantly different from the mean surrogate exponent). The other signals, measured in seismic areas, present magnitude σ value above the threshold in most cases; while the sign σ values above the threshold and those below are almost identically distributed. From this result, the magnitude series conveys information about the nonlinearity of the process underlying the signal variation. The signals measured in aseismic areas, like Laterza, are characterized by linear dynamics, while those measured in seismic areas by nonlinear dynamics. But, a better discrimination between both classes of signals can be obtained plotting all the points representing the original data and the surrogate data in the d_{mag}-d_{sig} plane. Fig. 8.12 shows this relation for the original signals (circle) and surrogates (crosses) of Tito (a), Giuliano (b), Marsico (c) and Laterza (d).

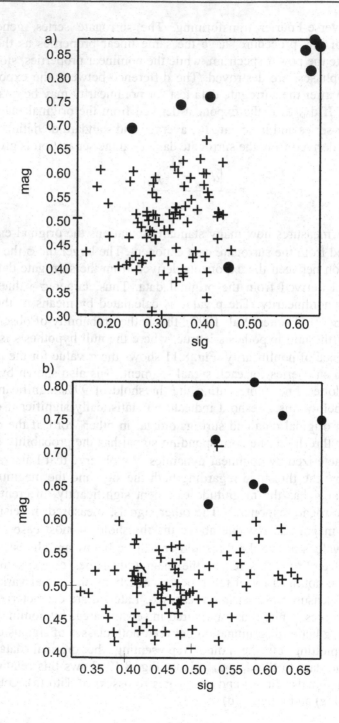

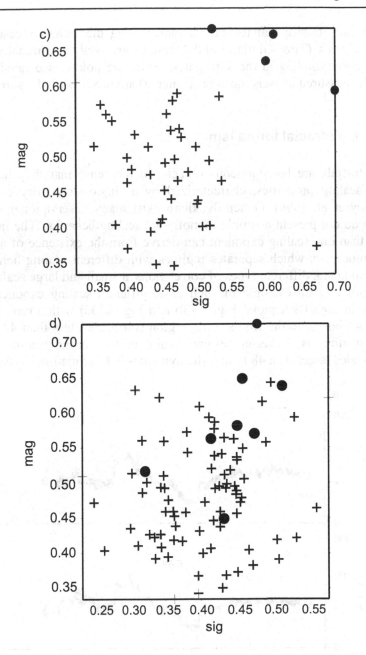

Fig. 8.12 d_{mag}-d_{sig} plane representation of the signal segments along with their surrogates: (a) Tito, (b) Giuliano, (c) Marsico and (d) Laterza

We can observe that the points representing the signals measured in seismic areas (Tito, Giuliano and Marsico) are well discriminated from those corresponding to the surrogates, while the points associated to the signals measured in aseismic areas (Laterza) are mixed with the surrogates.

8.7 Multifractal formalism

Monofractals are homogeneous objects, in the sense that they have the same scaling properties, characterized by a single singularity exponent (Stanley et al. 1996). Generally, there exist many observational signals, which do not present a simple monofractal scaling behavior. The need for more than one scaling exponent can derive from the existence of a crossover timescale, which separates regimes with different scaling behaviors, suggesting e.g. different types of correlations at small and large scales. Fig. 8.13 provides an example of such case: different scaling exponents are visible in both DFA plots (Fig. 8.13b and Fig. 8.13c) with a crossover at about 48 hours. In the first scaling region (timescales less than 48 hours) the dynamics is flicker-noise-type, while in the second scaling region (timescales larger than 48 hours) the dynamics is Brownian-noise-type.

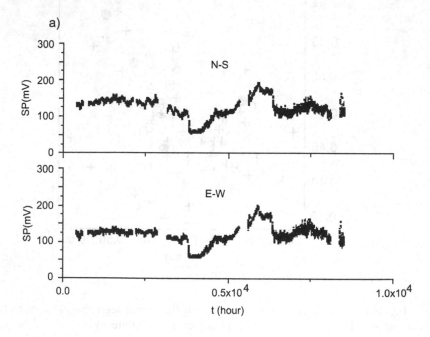

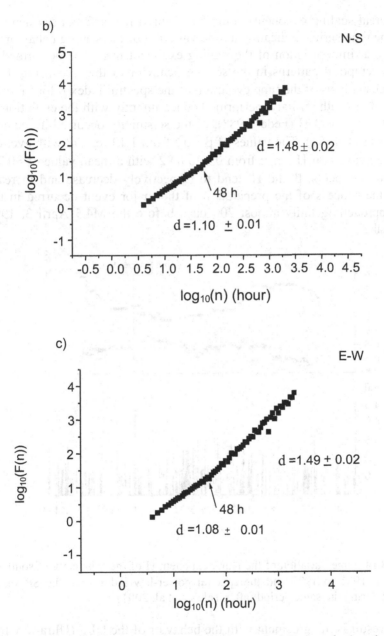

Fig. 8.13(a) Hourly SP means recorded at station Tramutola (southern Italy) during the 1995; (b and c) DFA results: clear power-law behaviour with scaling index d≈1.1 and d≈1.4 with a crossover at about 48 hours (after Telesca et al. 2002)

Different scaling exponents could be required for different segments of the same time series, indicating a time variation of the scaling behavior. In this case a time variation of the scaling exponent has to be performed to identify temporal patterns in the scaling behavior of the signal. Fig. 8.14 shows an analysis of the time evolution of the spectral index β for SP data, measured in southern Italy, performed concomitantly with the evolution of the Hurst exponent H (Feder 1988), of the seismicity occurred in the area (Telesca et al. 2001). The values of β vary from 1.12 to 1.78 with average ≈1.4. The values of H range from 0.5 to 0.92 with a mean value of ≈0.74. The two exponents, β and H, tend to respectively decrease and increase during the process of the preparation of the major event occurred in the area, approaching unity almost 200 days before the M4.5 April 3, 1996 earthquake.

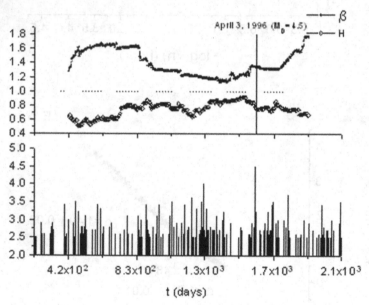

Fig. 8.14 Time variation of the Hurst exponent H of the seismicity of southern Italy from 1991 to 1997 and the spectral power-law index β of the SP signal recorded during the same period (after Telesca et al. 2001)

This result is in agreement with the behavior of the ULF (Ultra-low frequency) spectral exponent observed in Hayakawa (1999). The decrease of β and the increase of H both toward unity before the occurrence of the major event can be viewed as an indicator of self-organized criticality of the geophysical system governing both the SP and the seismic temporal fluctuations.

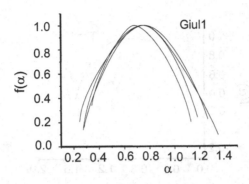

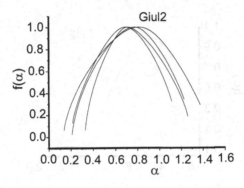

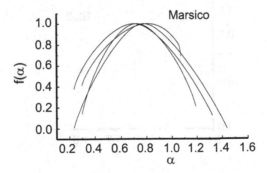

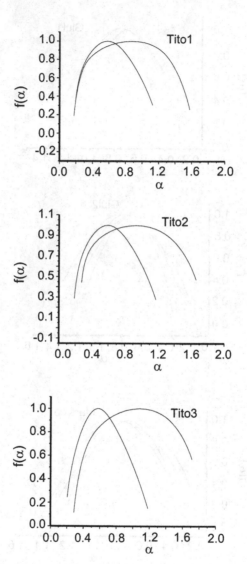

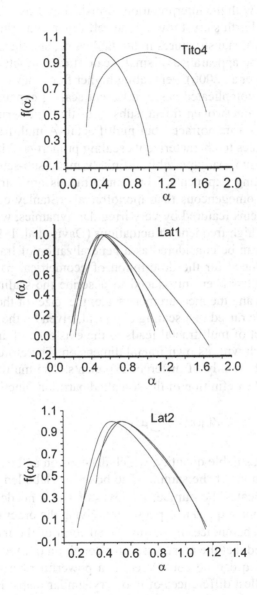

Fig. 8.15 Legendre spectra of the signals measured in southern Italy. For each signal, we selected 2 to 4 longest segments without gaps, whose length is approximately 10^3 points. All the spectra evidence the single-humped shape, typical of multifractal signals (after Telesca et al. 2004a)

In agreement with the interpretation furnished by Hayakawa (1999), the evolution of the Earth's crust toward the self-organized criticality involves the formation of fractal structures in the fault zone, and the decrease of β is consistent with the appearance of small-scale fractal structures in the focal zone (Hayakawa et al. 2000), generating higher frequency components.

In even more complicated cases, different scaling exponents can be revealed for many interwoven fractal subsets of the time series; in this case the process is not a monofractal but multifractal. A multifractal object requires many indices to characterize its scaling properties. Multifractals can be decomposed into many-possibly infinitely many-sub-sets characterized by different scaling exponents. Thus multifractals are intrinsically more complex and inhomogeneous than monofractals (Stanley et al. 1999) and characterize systems featured by very irregular dynamics, with sudden and intense bursts of high frequency fluctuations (Davis et al. 1994).

Multifractals can be considered as a generalization of fractal geometry, essentially developed for the description of geometrical patterns. Indeed, fractal geometry has been introduced to describe the scaling relationship between patterns and the measurement scale: the 'size' of the fractal object varies as the scale raised to a scaling exponent, given by the fractal dimension. The concept of multifractal leads to the existence of an infinite hierarchy of sets, each with its own fractal dimension. Therefore, multifractals require an infinite family of different exponents. The multifractal formalism is based on the definition of the so-called partition function $Z(q,\varepsilon)$,

$$Z(q,\varepsilon)= \sum_{i=1}^{N_{boxes}(\varepsilon)}\mu_i^q(\varepsilon). \tag{8.17}$$

The $\mu_i(\varepsilon)$ is a measurable quantity, which depends on ε, the size or scale of the boxes used to cover the sample. The boxes are labelled by the index i and $N_{boxes}(\varepsilon)$ indicates the number of boxes of size ε needed to cover the sample. The exponent q is a real parameter, giving the order of the moment of the measure. The choice of the functional form of the measure $\mu_i(\varepsilon)$ is arbitrary, provided that the most restrictive condition $\mu_i(\varepsilon) \geq 0$ is satisfied.

The parameter q can be considered as a powerful microscope, able to enhance the smallest differences of two very similar maps. Furthermore, q represents a selective parameter: high values of q enhance boxes with relatively high values for $\mu_i(\varepsilon)$; while low values of q favour boxes with relatively low values of $\mu_i(\varepsilon)$. The box size ε can be considered as a filter, so that big values of the size are equivalent to apply a large scale filter to the map (Diego et al. 1999). Changing the size ε, one explores the sample at different scales. Therefore, the partition function $Z(q, \varepsilon)$ furnishes information at different scales and moments.

The generalized dimension are defined by the following equation

$$D(q) = \lim_{\varepsilon \to 0} \frac{1}{q-1} \frac{\ln Z(q, \varepsilon)}{\ln \varepsilon} \qquad (8.18)$$

$D(0)$ is the capacity dimension; $D(1)$ is the information dimension, and $D(2)$ is the correlation dimension. An object is called monofractal if $D(q)$ is constant for all values of q, otherwise is called multifractal. In most practical applications the limit in Eq. (8.18) cannot be calculated, because we do not have information at small scales, or because below a minimum physical length no scaling can exist at all (Diego et al. 1999). Generally, a scaling region is found, where a power-law can be fitted to the partition function, which in that scaling range behaves as

$$Z(q, \varepsilon) \propto \varepsilon^{\tau(q)} \qquad (8.19)$$

The slope $\tau(q)$ is related to the generalized dimension by the following equation:

$$\tau = (q-1)D(q) \qquad (8.20)$$

A usual measure in characterizing multifractals is given by the singularity spectrum or Legendre spectrum $f(\alpha)$ that is defined as follows. If for a certain box j the measure scales as

$$\mu_j(\varepsilon) \propto \varepsilon^{\alpha_j} \qquad (8.21)$$

the exponent α, which depends upon the box j, is called Hölder exponent. If all boxes have the same scaling with the same exponent α, the sample is monofractal. The multifractal is given if different boxes scale with different exponents α, corresponding to different strength of the measure. Denoting as S_α the subset formed by the boxes with the same value of α, and indicating as $N_\alpha(\varepsilon)$ the cardinality of S_α, for a multifractal the following relation holds:

$$N_\alpha(\varepsilon) \propto \varepsilon^{-f(\alpha)} \qquad (8.22)$$

By means of the Legendre transform the quantities α and $f(\alpha)$ can be related with q and $\tau(q)$:

$$\alpha(q) = \frac{d\tau(q)}{dq} \qquad (8.23)$$

$$f(\alpha) = q\alpha(q) - \tau(q)$$
(8.24)

The curve $f(\alpha)$ is a single-humped function for a multifractal, while reduces to a point for a monofractal. In order to quantitatively recognize possible differences in Legendre spectra stemming from different signals, it is possible to fit, by a least square method, the spectra to a quadratic function around the position of their maxima at α_0 (Shimizu et al. 2002):

$$f(\alpha) = A(\alpha - \alpha_0)^2 + B(\alpha - \alpha_0) + C$$
(8.25)

where parameter B serves as an asymmetry parameter, which is zero for symmetric shapes, positive or negative for a left- or right-skewed (centred) shape, respectively. Another parameter is the width of the spectrum, that estimates the range of α where $f(\alpha)>0$, obtained extrapolating the fitted curve to zero; thus the width is defined as

$$W = \alpha_1 - \alpha_2$$

where

$$f(\alpha_2) = f(\alpha_1) = 0$$
(8.27)

The width W measures the length of the range of fractal exponents in the signal; therefore, the wider the range, the "richer" the signal in structure. The asymmetry parameter B provides about the dominance of low or high fractal exponents λ with respect to the other.

As an example of application of the multifractal method to reveal dynamical features in SP data, we discuss the multifractal discrimination between the nine SP signals considered in the previous section, and recorded in seismic (Giul1, Giul2, Marsico, Tito1, Tito2, Tito3 and Tito4) and aseismic (Lat1 and Lat2) areas in southern Italy. Fig. 8.15 shows the Legendre spectra for the selected segments for each signal. All the spectra present the typical single-humped shape that characterizes multifractal signals. Tito signals present a couple of curves very different from each other, the wider corresponding to the segment of data extracted in the temporal range in which an increase of seismic activity has been recorded (see Fig. 8.9). For each multifractal spectrum the three parameters, maximum α_0, asymmetry B and width W have been calculated; then an average among them has been performed, obtaining a set of three multifractal parameters for each original signal. In order to evaluate the significance of the results, for each segment 10 surrogate series have been generated, and for each of them, the multifractal spectrum and the three multifractal parameters. The maximum-width, maximum-asymmetry and asymmetry-width relations for the original and the surrogate series are described respectively in details by Telesca et al.(2004a). The SP signals measured in seismic areas (Giuliano, Tito and Marsico) are discriminated from the surrogate series; while Laterza signals do not show any separation between the points representing the original signals and the those representing the surrogates. Fig. 8.16 shows a 3D plot of the multifractal parameters, where the signals Lat1 and Lat2, recorded in aseismic area, are well discriminated from the other signals measured in seismic areas.

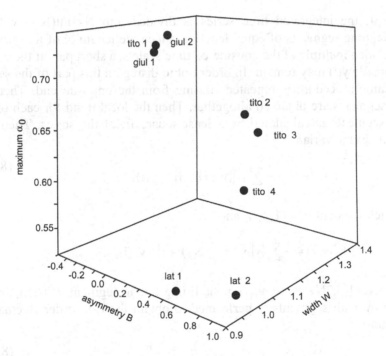

Fig. 8.16 3D plot of the multifractal parameters: the signals Lat1 and Lat2, recorded in aseismic area, are well discriminated from the other signals measured in seismic areas (after Telesca et al. 2004b)

8.8 Multifractal detrended fluctuation analysis

Observational data often present clear irregular dynamics, characterized by sudden bursts of high frequency fluctuations, which suggest performing a multifractal analysis evidencing the presence of different scaling behaviours for different intensities of fluctuations. Furthermore, the signal may appear nonstationary, and, for this reason, the Multifractal Detrended Fluctuation Analysis (MF-DFA) (Kantelhardt et al. 2002) could be a useful tool to characterize multifractality.

The method is based on the conventional DFA. Thus, it operates on the time series x(i), where i=1,2,...,N and N is the length of the series. Assume that x(i) are increments of a random walk process around the average x_{ave}, the "trajectory" or "profile" is given by the integration of the signal

$$y(i) = \sum_{k=1}^{i} [x(k) - x_{ave}]$$

Next, the integrated time series is divided into N_S=int(N/s) without overlapping segments of equal length s. Since the length N of the series is often not a multiple of the considered time scale s, a short part at the end of the profile y(i) may remain. In order not to disregard this part of the series, the same procedure is repeated starting from the opposite end. Thereby, $2N_S$ segments are obtained altogether. Then the local trend for each of the $2N_S$ segments is calculated by a least square fit of the series. Then one calculates the variance

$$F^2(s,v) = \frac{1}{s} \sum_{i=1}^{s} \{y[(v-1)s+i] - y_v(i)\}^2 \qquad (8.28)$$

for each segment v, $v=1,..,N_S$ and

$$F^2(s,v) = \frac{1}{s} \sum_{i=1}^{s} \{y[N-(v-N_S)s+i] - y_v(i)\}^2 \qquad (8.29)$$

for $v=N_S+1,..,2N_S$. Here, $y_v(i)$ is the fitting line in segment v. Then, an average over all segments is performed to obtain the q-th order fluctuation function

$$F_q(s) = \left\{ \frac{1}{2N_S} \sum_{v=1}^{2N_S} [F^2(s,v)]^{\frac{q}{2}} \right\}^{\frac{1}{q}} \qquad (8.30)$$

where, in general, the index variable q can take any real value except zero. The value h(0) corresponds to the limit h(q) for q→0, and cannot be determined directly using the averaging procedure of Eq. 8.30 because of the diverging exponent. Instead, a logarithmic averaging procedure has to be employed,

$$F_0(s) \equiv \exp\left\{ \frac{1}{4N_S} \sum_{v=1}^{2N_S} \ln[F^2(s,v)] \right\} \approx s^{h(0)} \qquad (8.31)$$

Repeating the procedure described above, for several time scales s, $F_q(s)$ will increase with increasing s. Then analyzing log-log plots $F_q(s)$ versus s for each value of q, the scaling behaviour of the fluctuation functions can be determined. If the series x_i is long-range power-law correlated, $F_q(s)$ increases for large values of s as a power-law:

$$F_q(s) \propto s^{h(q)} \qquad (8.32)$$

In general the exponent h(q) will depend on q. For stationary time series, h(2) is the well defined Hurts exponent H (Feder 1988). Thus, we call h(q) the generalized Hurst exponent. Monofractal time series are characterized by h(q) independent of q. The different scaling of small and large fluctuations will yield a significant dependence of h(q) on q. For positive q, the segments v with large variance (i.e. large deviation from the corresponding fit) will dominate the average $F_q(s)$. Therefore, if q is positive, h(q) describes the scaling behaviour of the segments with large fluctuations; and generally, large fluctuations are characterized by a smaller scaling exponent h(q) for multifractal time series. For negative q, the segments v with small variance will dominate the average Fq(s). Thus, for negative q values, the scaling exponent h(q) describes the scaling behaviour of segments with small fluctuations, usually characterized by larger scaling exponents.

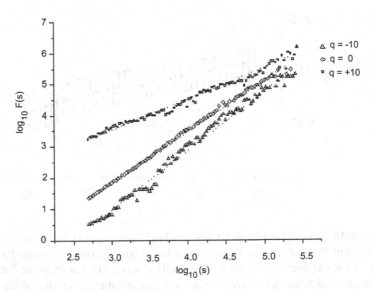

Fig. 8.17 Fluctuation functions for q=-10, 0, +10 and time scales ranging between $5 \cdot 10^3$ min to N/4, where N is the total length of the series

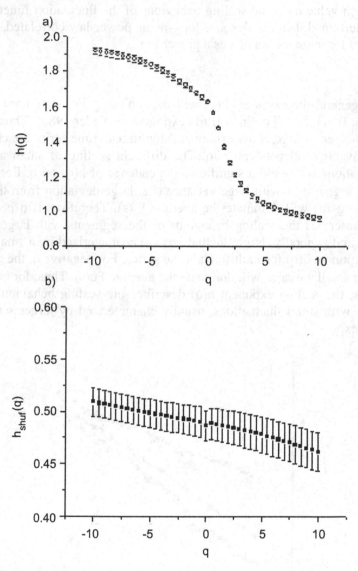

Fig. 8.18 Generalized Hurst exponents h(q) for q varying between –10 and 10. It is evident the multifractal behaviour of the original series (a); while the smaller range of the exponents concerning the shuffled series (10 realizations; the average ± 1 standard deviation is shown) indicates that the multifractality of the original signal depends mostly on the different long-range correlations for small and large fluctuations (b)

Two types of multifractality that underlies the q-dependence of the generalized Hurst exponent in time series can be discriminated: (i) due to a broad probability density function for the values of the time series, and (ii) due to different long-range correlations for small and large fluctuations. Both of them need a multitude of scaling exponents for small and large fluctuations. The easiest way to discriminate between these two types of multifractality is by analyzing the corresponding randomly shuffled series. In the shuffling procedure the values are put into random order, and although all correlations are destroyed, the probability density function remains unchanged. Hence the shuffled series coming from multifractals of type (ii) will exhibit simple random behaviour with $h_{shuf}(q)=0.5$, which corresponds to purely random dynamics. While those coming from multifractals of type (i) will show $h(q)=h_{shuf}(q)$, since the multifractality depends on the probability density. If both types of multifractality characterize the time series, thus the shuffled series will show weaker multifractality than the original one. Fig. 8.17 provides three fluctuation functions $F_q(s)$ (q=-10, 0, 10) for a self-potential signal measured in southern Italy for time-scales s ranging from $5 \cdot 10^2$ min to N/4, where N is the total length of the series. The length of the series ($N \sim 1.2 \cdot 10^6$) allows us to consider the estimated exponents reliable. Fig.8.18 shows the q-dependence of the generalized Hurst exponent h(q) determined by fits in the regime $5 \cdot 10^2$ min < s < N/4. Also shown are the generalized Hurst exponents versus q, averaged over 10 randomly shuffled versions of the original time series. The error bars delimit the 1-σ range around the mean values. The $h_{shuf}(q)$-values range around 0.5, but with a slight q-dependence; this indicates that the most multifractality of the self-potential data is due to different long-range correlations for small and large fluctuations.

8.9 Discussion

The geophysical phenomenon underlying the geoelectrical variations is complex and the use of fractal analysis to detect scaling laws in the statistics describing the geoelectrical time series can lead to a better understanding of the physics of the process. Recently, the study of the dynamics of extended systems have suggested that complex systems are very common in nature and a typical effect in the time domain is known as $1/f^\alpha$ noise, so the power spectra of these processes exhibit a linear behaviour on log-log scales. Monofractals and multifractals are particularly well suited for characterizing long-range power-law correlated self-potential signals. The mono- and multifractality observed in self-potential signals recorded in seismic areas can reflect the irregularity and heterogeneity of the crust, within which phenomena generating self-potential fields occur. Therefore the structure of the self-potential signal is linked to the structure of the seismic focal zone. In fact, the geometry and the structure of individual fault zones can be represented by a network with an anisotropic distribution of fracture orientations, and consisting of fault-related structures including small faults, fractures, veins and folds. This is a consequence of the roughness of the boundaries between each component and the interaction between the distinct components within the fault zone (O'Brien et al. 2003). In fact earthquake faulting is characterized by irregular rupture propagation and non-uniform distributions of rupture velocity, stress drop and co-seismic slip. These observations indicate a non-uniform distribution of strength in the fault zone, whose geometry and mechanical heterogeneities are important factors to be considered in the prediction of strong motion. Experimental studies on the hierarchical nature of the processes underlying fault rupture, leading to the possibility of recognizing the final preparation stage before a large earthquake have been performed (Lei et al. 2003). The phenomenon of electrofiltration or streaming potential, which is known to be the most relevant phenomenon that could originate the geoelectrical field, can be influenced by the structure of the seismic focal zone, under the condition of dilatancy-induced crack formation and propagation. The fractal analysis of this generated geoelectrical signal could reveal the fractality of such structure.

Acknowledgements

We thank V.P. Dimri for his invitation to write this review. Our thanks are due to World Scientific Publishing Co. Pte. Ltd. for reproduction of figures 8.2 and 8.13, Blackwell Publishing Ltd for figures 8.3 and 8.4, Elsevier publishers for figure 8.6, 8.9 and 8.15; AGU Publishers for figure 8.14; and Editrice Compositori publisher for figure 8.16.

8.10 References

Ashkenazy Y, Ivanov PC, Havlin S, Peng C-H, Goldberger AL, Stanley HE (2001) Magnitude and sign correlations in heartbeat fluctuations. Phys Rev Lett 86: 1900-1903

Ashkenazy Y, Havlin S, Ivanov PC, Peng C-K, Schulte-Frohlinde V, Stanley HE (2003) Magnitude and sign scaling in power-law correlated time series. Physica A 323: 19-41

Berry MV (1979) Diffractals. J Phys A Math Gentile 12: 781-792

Buldyrev SV, Goldberger AL, Havlin S, Peng C-K, Stanley HE (1994) Fractals in biology and medicine: From DNA to the Heartbeat. In: Bunde A, Havlin S (eds) Fractals in Science, Springer-Verlag, Berlin Heidelberg New York, pp 48-87

Burlaga LF, Klein LW (1986) Fractal structure of the interplanetary magnetic field. J Geophys Res 91: 347

Brace WF, Paulding Jr BW, Scholz CH (1966) Dilatancy in the fracture of cristal-line rocks. J Geophys Res 71: 3939-3953

Box GEP, Jenkins GM (1976) Time Series Analysis. Holden-Day, S. Francisco

Colangelo G, Lapenna V, Telesca L (2003) Analysis of correlation properties in geoelectrical data. Fractals 11: 27-38

Corwin RF, Hoover DB (1979) The SP method in geothermal exploration. Geophysics 44: 226-245

Cuomo V, Lapenna V, Macchiato M, Serio C, Telesca L (1999) Stochastic behaviour and scaling laws in geoelectrical signals measured in a seismic area of southern Italy. Geophys J Int 139: 889-894

Davis A, Marshak A, Wiscombe W (1994) Wavelet-based multifractal analysis of non-stationary and/or intermittent geophysical signals. In: Foufoula-Georgiou E, Kumar P (eds) Wavelets in geophysics: Academic Press, New York, pp 249-298

Di Bello G, Heinicke J, Koch U, Lapenna V, Macchiato M, Martinelli G, Piscitelli S (1998) Geophysical and geochemical parameters jointly monitored in a seismic area of Southern Apennines (Italy). Phys Chem Earth 23: 909-914

Diego JM, Martinez-Gonzales E, Sanz JL, Mollerach S, Mart VJ (1999) Partition function based analysis of cosmic microwave background maps. Mon Not R Astron Soc 306: 427-436

Di Maio R, Patella D (1991) Basic theory of electrokinetic effects associated with earthquake. Boll Geof Teor Appl 33: 130-131

Di Maio R, Mauriello P, Patella D, Petrillo Z, Piscitelli S, Siniscalchi A, Veneruso M (1997) Self-potential, geoelectric and magnetotelluric studies in Italian active volcanic areas. Annali di Geofisica 40: 519-537

Dimri VP, Ravi Prakash M (2001) Scaling of power spectrum of extinction events in the fossil record. Earth Planet Sci Lett 186: 363-370

Dobrovolsky, IP (1993) Analysis of preparation of a strong tectonic earthquake. Phys Solid Earth 28: 481-492

Dobrovolsky IP, Zubkov SI, Miachkin VI (1979) Estimation of the size of earthquake preparation zones. Pure Appl Geoph 117: 1025-1044

Kantelhardt JW, Koscienly-Bunde E, Rego HHA, Havlin S, Bunde A (2001) Detecting long-range correlations with detrended fluctuation analysis. Physica A 295: 441-454

Kantelhardt JW, Zschiegner SA, Konscienly-Bunde E, Havlin S, Bunde A, Stanley HE (2002) Multifractal detrended fluctuation analysis of nonstationary time series. Physica A 316: 87

Keller GV, Frischknecht FC (1966) Electrical Methods in Geophysical Prospecting, Pergamon Press, Oxford

Havlin S, Amaral LAN, Ashkenazy Y, Goldberger AL, Ivanov PCh, Peng C-K, Stanley HE (1999) Application of statistical physics to heartbeat diagnosis. Physica A 274: 99-110

Hayakawa M (1994) Direction finding of seismogenic emissions. In: Hayakawa M, Fujinawa Y, (eds). Electromagnetic phenomena related to earthquake prediction. Terra Sci. Pub. Co., Tokyo, pp 493-494

Hayakawa M, Hattori K, Itoh T, Yumoto K (2000) ULF electromagnetic precursors for an earthquake at Biak, Indonesia on February 17, 1996. Geophys Res Lett 27: 1531-1534

Hayakawa M, Ito T, Smirnova N (1999) Fractal analysis of ULF geomagnetic data associated with the Guam earthquake on August 8, 1993. Geophys Res Lett 26: 2797-2800

Higuchi T (1988) Approach to an irregular time series on the basis of the fractal theory. Physica D 31: 277-283

Higuchi T (1990) Relationship between the fractal dimension and the power law index for a time series: a numerical investigation. Physica D 46: 254-264

Feder J (1988) Fractals, Plenum Press, New York

Lapenna V, Patella D, Piscitelli S (2000) Tomographic analysis of SP data in a seismic area of Southern Italy. Annali di Geofisica 43: 361-373

Lei X, Kusunose K, Nishizawa O, Satoh T (2003) The hierarchical rupture process of a fault: an experimental study. Phys Earth Planet Int 137: 213

Lomb NR (1976) Least-squares frequency analysis of unequally spaced data. Astrophyisics and Space Science 39: 447-462

Martinelli G, Albarello D (1997) Main constraints for siting monitoring networks devoted to the study of earthquake related hydrogeochemical phenomena in Italy. Annali di Geofisica 40: 1505-1522

Mizutani H, Ishido T, Yokokura T, Ohnishi S (1976) Electrokinetic phenomena associated with earthquakes. Geophys Res Lett 3: 365-368

Nur A (1972) Dilatancy pore fluids and premonitory variations of ts/tp travel times. Bull Seism Soc Am 62: 1217-1222

O'Brien GS, Bean CJ, McDermott F (2003) A numerical study of passive transport through fault zones. Earth Planet Sci Lett 214: 633-643

Ogilvy AA, Ayed MA, Bogoslovsky VA (1969) Geophysical studies of water leakages from reservoirs. Geophys Prosp 22: 36-62

Parasnis DS (1986) Principles of Applied Geophysics. Chapman and Hall, London/New York

Park SK (1997) Monitoring resistivity changes in Parkfield, California 1988-1995. J Geophy Res 102: 24545-24559

Patella D (1997) Introduction to ground surface SP tomography. Geophys Prospect 45: 653-681

Peng C.-K, Havlin S, Stanley HE, Goldberger AL (1995) Quantification of scaling exponents and crossover phenomena in nonstationary heartbeat time series. CHAOS 5: 82-87

Rikitake T(1988) Earthquake prediction: an empirical approach. Tectonophysics 148: 195-210

Scholz CH (1990) The mechanics of earthquakes and faulting. Cambridge University Press, Cambridge

Sharma PS (1997) Environmental and engineering geophysics. Cambridge University Press, Cambridge

Shimizu Y, Thurner S, Ehrenberger K (2002) Multifractal spectra as a measure of complexity in human posture. Fractals10:103-116

Stanley HE, Afanasyev V, Amaral LAN, Buldyrev SV, Goldberger AL, Havlin S, and Leschhorn H, Maass P, Mantegna RN, Peng C.-K., Prince PA, Salinger RA, Stanley MHR, Viswanathan GM (1996) Anomalous fluctuations in the dynamics of complex systems: from DNA and physiology to econophysics. Physica A 224: 302-321

Stanley HE, Amaral LAN, Goldberger AL, Havlin S, Ivanov PCh, Peng C-K (1999) Statistical physics and physiology: Monofractal and multifractal approaches. Physica A 270: 309-324

Telesca L, Cuomo V, Lapenna V, Macchiato M (2001) A new approach to investigate the correlation between geoelectrical time fluctuations and earthquakes in a seismic area of southern Italy. Geophys Res Lett 28: 4375-4378

Telesca L, Lapenna V, Macchiato M (2002) Fluctuation analysis of the hourly time variability in observational geoelectrical signals. Fluctuation Noise Lett 2: L235-L242

Telesca L, Colangelo G, Lapenna V, Macchiato M (2003) Monofractal and multifractal characterization of geoelectrical signals measured in southern Italy. Chaos Solitons & Fractals 18: 385-399

Telesca L, Balasco M, Colangelo G, Lapenna V, Macchiato M (2004a). Investigating the multifractal properties of geoelectrical signals measured in southern Italy. Phys Chem Earth 29: 295-303

Telesca L, Colangelo G, Lapenna V, Macchiato M (2004b) A preliminary study of the site-dependence of the multifractal features of geoelectric measurements. Annals of Geophysics 47: 11-20

Theiler J, Eubank S, Longtin A, Galdrikian B, Farmer JD (1992) Testing for nonlinearity in time series: the method of surrogate data. Physica D 58: 77-94

Thurner S, Lowen SB, Feurstein MC,Heneghan C, Feichtinger HC, Teich MC (1997) Analysis, Synthesis, and Estimation of Fractal-Rate Stochastic Point Processes. Fractals 5: 565-596

Tramutoli V, Di Bello G, Pergola N, Piscitelli S (2001) Robust satellite techniques for remote sensing of seismically active areas. Annali di Geofisica 44: 295-312

Troyan VN, Smirnova NA, Kopytenko YA, Peterson T, Hayakawa M (1999) Development of a complex approach for searching and investigation of electromagnetic precursors of earthquakes: Organization of experiments and analysis procedures. In: Hayakawa, M., (ed) Atmospheric and ionospheric electromagnetic phenomena associated with earthquakes, Terra Sci. Pub. Co., Tokyo, pp 147-170

Vallianatos F, Tzanis A (1999) On possible scaling laws between Electric Earthquake Precursors (EEP) and Earthquake Magnitude. Geophys Res Lett 26: 2013-2016

Zhao Y, Qian F (1994) Geoelectric precursors to strong earthquakes in China Tectonophysics 233: 99-113

Chapter 9. Earth System Modeling Through Chaos

H. N. Srivastava

Formerly in India Meteorological Department, New-Delhi, India

9.1 Summary

Modeling parameters using deterministic chaos have been discussed for Earth system through atmospheric pressure, maximum and minimum temperature, monsoon rainfall, cyclonic storm tracks, long term climate, ozone, radio refractive index, magnetosphere ionosphere system, volcanoes, earthquakes and fluid flows in core and mantle using the method of Grassberger and Procaccia and Lyapunov exponents. It was found that the atmospheric phenomena generally showed a strange attractor dimension of 6 to 7 implying at least 7 to 8 parameters for modeling the system. On the other hand, the magnetosphere-ionosphere system had a low dimension. Most interesting results were found for earthquakes whose strange attractor dimension provides a methodology for differences between interplate and intraplate Indian region. It also provides a dynamical justification for delineation of seismicity patterns based on epicenters of earthquakes on different closely located fault systems up to 500 km radius from the impending earthquake.

Another interesting result pertains to the Koyna region, India where a low strange attractor dimension of 4.5 provides justification for earthquake predictability programme in this region.

9.2 Introduction

Earth as a dynamic system involves extensive studies from the core and mantle processes, tectonic plate movements (causing earthquakes and volcanoes), atmospheric, ocean solar and terrestrial relationships. The predominant effects on the magnetosphere and ionosphere have been attributed to the solar radiation and particle fluxes. The atmosphere is affected by the solar irradiance. The plate motions are governed by the convection processes in the mantle. Thus the interactions and feed backs among the

Earth system components are primarily responsible for complexity in the phenomenon which occurs in the biosphere, atmosphere, hydrosphere, lithosphere, mantle and the core. A unified approach is therefore required to model these phenomenon which cause global changes. In view of the problems to solve the complicated dynamical equations, recourse is taken through chaotic dynamics.

9.3 Methodology

In order to examine whether a system is chaotic, methods like Grassberger and Procaccia (1983), Lyapunov exponents (Wolf et al. 1985), are widely used.

9.3.1 Strange attractor dimension

In the Grassberger and Procaccia (1983) method, phase space of the dynamical system is reconstructed by using an observed time series of number of earthquakes or any other parameter as variables and the time series of the same variable but shifted by $(n-1)$ time lags,

$$X(t) = [x(t), x(t+1), \ldots\ldots\ldots\ldots x(t+n-1)]$$

A set of N points on an attractor embedded in a phase space of n dimensions is obtained from the time series given by,

$$x_i((t_i), \ x(t_i + \tau) \ldots\ldots\ldots\ldots\ldots\ldots x(t_I+(n-1) \tau) \tag{9.1}$$

which stands for a point of phase space, the difference $(X_I - X_j)$ from the N-1 remaining points are computed.

The correlation function of the attractor $C_m(r)$ is given by

$$C_m (r) = \frac{1}{N^2} \sum_{i,j=1}^{N} H[r - (X_i - Xj)], i \neq j \tag{9.2}$$

for embedding dimension m and where H is Heaviside function i.e.

$$H(x)=0 \ , \ \text{if } x<0$$
$$H(x)=1 \ , \ \text{if } x>0 \tag{9.3}$$

and r is distance

The dimensionality d of the attractor is related to $C_m(r)$ by the relation

$$C_m(r) = r^d \text{ or } C_m(r) = d \ln(r) \tag{9.4}$$

The dimensionality d of the attractor is given by the slope of the $C_m(r)$ versus $\ln(r)$. The slope of the scaling region is obtained for various embedding dimensions. As we increase the embedding dimension m, the slope saturates to a limiting value obtained for two consecutive delay times. This gives the strange attractor dimension. The minimum and the maximum number of parameters for predictability are d + 1 and 2d respectively.

9.3.2 Lyapunov exponent

We consider a system described by N ordinary differential equations

$$\frac{dx_i}{dt} = F_i(x_1, x_2, \ldots x_n), i = 1, 2, \ldots N \tag{9.5}$$

The solution space for this problem conceptually follows the solutions that start within a hypersphere of radius r. As the solution evolves, the hypersphere is deformed into a hyperellipsoid with principal axis $E_i(t)$. The Lyapunov Exponent is given by,

$$\lambda_i = x[1 / t(E_i(t) / r)] \tag{9.6}$$
$$\lim_{t \to \infty, r \to 0}$$

If all $\lambda_l \leq 0$, all solutions that start with initial conditions close to each other will converge i.e., there is no sensitivity to initial conditions. But if just one λ_i is positive, the nearby solutions will diverge, i.e., there will be extreme sensitivity to the initial conditions. The growth in uncertainty in time t is given by

$$N = N_0 e^{\lambda t} \tag{9.7}$$

where N_0 is the initial condition and λ is related to the concept of entropy in information theory and also related to another concept i.e., the Lyapunov exponent, which measures the rate at which the nearby trajectories of a system in phase space diverge.

It may be noted that in order to avoid spurious results being obtained, the number of earthquakes N should satisfy the Ruelles criterion

$$2 \log_{10} N > D \tag{9.8}$$

where D is the strange attractor dimension.

Some significant results pertaining to Earth system from core of the Earth to the magnetosphere through the intervening ocean-atmospheric region would throw light about the predictability of the phenomenon which affects our lives in one way or the other.

9.4 Earth

Earth consists of three main layers called the crust, mantle and the core. Several text books (e.g. Gutenberg 1959, Lowrie 1997) give detailed description of these layers.

9.4.1 Outer core

Rikitake (1958) first parameterized and solved a set of dynamic equations which were shown to be examples of deterministic chaos. This was based on the characteristics of the core dynamics due to which the Earth's magnetic field undergoes reversals and is expected to be turbulent.

9.4.2 Earth's mantle

The heat transport in the Earth's mantle is mainly attributed to thermal convection. Stewart and Turcotte (1989) considered a higher order expansion in Fourier series based on Lorenz equations and showed that the time evolution of the solution is fully chaotic for Rayleigh Number = 4.5×10^4. No fixed points were stable. The phenomenon has assumed greater importance because the movement of the global plates is governed by the heat convection in the mantle pushing the plates away at the oceanic ridges. Some of the volcanic eruptions also originate in this region and are reported to be chaotic. The deep focus earthquakes in the mantle (up to about 700 km) are, however, least destructive, whose dynamics is least understood.

9.4.3 Earth's crust

The most destructive earthquakes generally originate in the upper crustal layers. The earthquakes of magnitude 8 or more may occur near the boundaries of the global plates of continent-continent (collision) or continent oceanic (subduction) type. The largest magnitude of earthquakes is generally 6 to 6.5 near the mid oceanic ridges. However, the largest magnitude of the earthquakes may reach 8 in the transform faults such as San Andreas in California. Intraplate seismicity provides greater challenges to seismologists. The recurrence of a great earthquake in Bhuj, India in January 2001 only after a gap of 183 years within the Indian plate attracted the attention of seismologists throughout the world. Chaotic methods have been used to study the dynamic behaviour of crustal earthquakes, which occurred in different parts of the world.

San Andreas fault

California (U.S.A.) is considered to be one of the most seismically active regions of the world. The section of San Andreas fault near Parkfield is tectonically more interesting because the great 1857 earthquake in southern California was preceded by foreshocks with their epicenters near Parkfield. This region is characterized by 'characteristic earthquakes'. Closer examination of Parkfield earthquakes, however, suggests that different inter-event intervals of 12 and 32 years between 1922, 1934 and 1966 earthquakes are inconsistent with a simple application of either the time or slip predictable recurrence models.

   Horowitz (1989) and Julian (1990) reported that a strange attractor exists in the Parkfield region of California. However, Beltrami and Mareschal (1993) did not find any evidence of a chaotic process in the region through generation of a random series and using seismic energy release of earthquake. They suggested that the occurrence in this region is random or had a strange attractor dimension larger than 12 implying inherent limitations in evolving a predictive model. In view of these results of divergence nature, Srivastava and Sinha Ray (1999) re-examined earthquake predictability in this region using two different approaches namely, Lyapunov exponent and strange attractor dimension. This was based on 75000 earthquakes on which the analysis of Beltrami and Mareschal (1993) was based during 1969-1987. Fig. (9.1a) shows the plot of $\ln C_m(r)$ versus $\ln(r)$ for embedding dimensions from 2 to 16. Fig. 9.1(b) gives the strange attractor dimension as 6.3 since saturation occurred for time delays of 4 and 5 days. This suggested that at least 7 parameters are needed for the predictability of earthquakes in the region. This unequivocally lent support to the results reported by Horowitz (1989) who found a chaotic process in this region. Further evidence of chaotic process in the region was provided by small positive value of the largest Lyapunov exponent (0.045).

Indian region

   Bhattacharya and Srivastava (1992) found a strange attractor of dimension 6.9 in the Hindukush region.

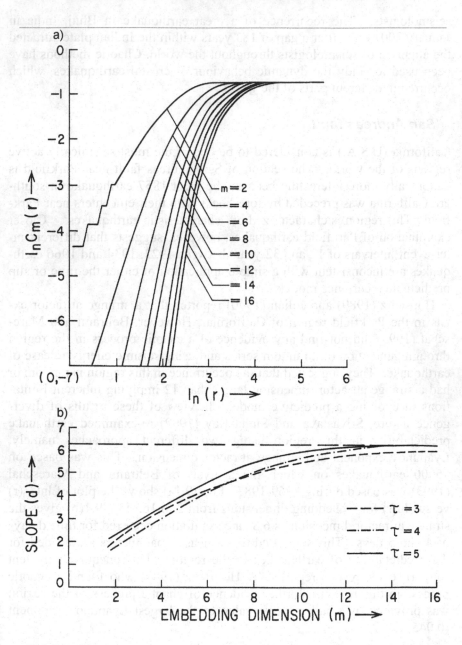

Fig. 9.1 (a) Distance dependence of the correlation function for a sequence of embedding dimensions in California region. (b) Dimensionality D of the attractor as a function of embedding dimensions, (after Srivastava and Sinha Roy 1999).

Tiwari et al. (2003) found that in the central India, K2 entropy had a low nonzero value (0.2 – 0.4) which suggested the occurrence of a weak "memory" and some predictability. In the northeast Indian region, Srivastava et al. (1996) found that a strange attractor of dimension 6.1 exists within 220 km and 440 km radius around Shillong with a positive value of the Lyapunov exponent. Using different data sets, however, Tiwari and Rao (2001) found a weak support to the chaotic process in this region.

In the peninsular India, Srivastava et al. (1994a) found a low strange attractor dimension of about 4.5 in the Koyna region. It was suggested as a new evidence of seismotectonics in the region. This value is considerably less as compared to the interplate Indian region. The value compares with that reported for the Aswan region of Egypt which is also characterized as a shield region. (Srivastava et al. 1995)

Japan region

Palvos et al. (1994) found a low strange attractor dimension in Japanese region using a micro earthquake catalogue which is also intraplate region.

9.4.4 Volcanic eruptions

The sequence of eruptions for a given volcano shows a complex temporal pattern. Many studies have brought out the possibility that dates of eruptions might exhibit "nonrandom" pattern which could be used in forecasting eruption. In this case, one is interested in the dynamics over very large time spans between 10 and 1000 years or larger. In order to characterize the volcano's dynamics, it would be necessary to record at least the time evolution of a physical variable over such long time intervals. Because of the relatively recent systematic physical study of volcanoes, such a program can be applied only to the variables which have been recorded over long time intervals, namely the times at which an eruption begins and ends. Using the data for 46 eruptions for the Kilauea and 38 for Mauna Loa volcanoes in Hawaii and by grouping both in a single set to maximize the data, the strange attractor dimension was found as 4.6 which was more than twice that for Piton de la Fournaise eruption (Sornette et al. 1991). It was surmised that each of the two volcanoes forming the complex structure of Hawaii, are weakly coupled with a dimension of about 2.

9.5 Atmosphere

9.5.1 Meteorological observations

Srivastava et al. (1994b) computed strange attractor dimension for daily sea level pressure, maximum temperature and daily minimum temperature for twelve well distributed stations over India following the method given by Grassberger and Procaccia (1983).

Table 9.1 Strange attractor dimension for meteorological parameters at Indian stations from 1969-1991

S.No	Station Name	Atmospheric Pressure (hp)	Max Temp. (^{0}C)	Min Temp. (^{0}C)
1	Thiruvananthapuram (TRV)	5.5	5.7	5.7
2	Madras (MDS)	5.8	5.7	5.9
3	Mangalore (MNG)	5.9	5.8	5.7
4	Visakhapatnam (VSK)	5.7	5.7	5.8
5	Hyderabad (HYD)	5.5	5.9	5.3
6	Pune (PNE)	5.7	5.3	5.1
7	Bombay (BMB)	5.7	5.3	5.7
8	Nagpur (NGP)	5.3	5.3	5.7
9	Calcutta (CAL)	5.7	5.7	5.7
10	Bhopal (BHP)	5.6	5.5	5.3
11	Gauhati (GHT)	5.3	5.3	5.1
12	Amritsar (AMR)	5.7	5.9	5.7

Daily data for 23 years (1969-1991) from India Meteorological department was utilized for this purpose. lnCm(r) was plotted against ln(r); the dimension d was calculated for embedding dimensions from 2 to 16 from the scaling region. The computation was made for delay time $\tau = 1$ day, $\tau = 2$ day and $\tau = 3$ day. Fig.9.2a shows the plot of lnCm(r) vs. ln(r) for delay time $\tau = 3$ day, (daily pressure of Nagpur). Fig. 9.2b shows plot of slope d against embedding dimension m. The saturation value of the fractal dimension was obtained at $\tau = 2$ and 3 day and for embedding dimension 14. Similar computations were made for all parameters for each station. The strange attractor dimension was found to vary between 5 and 6. The results are shown in Table 9.1. These are slightly smaller than the Berlin surface pressure, which gave the dimension of the attractors as 6.8 – 7.1 (Fraedrich 1987).

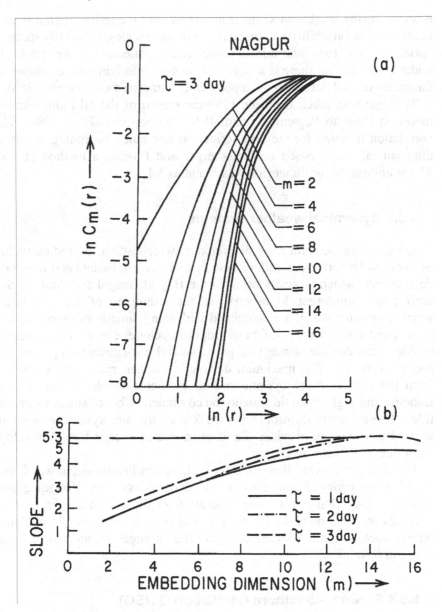

Fig. 9.2 Daily Pressure (1969-1991) (a) lnCm(r) Vs ln(r) (b) Slope D against embedding dimension m (after Srivastava and Sinha Roy 1997)

9.5.2 Monsoon rainfall

Rainfall during southwest summer monsoon over the Indian region is well known for its variability on a wide range of time scales, especially its large variability on the inter-annual time scale. Attempts were made to understand whether there is a strange attractor underlying the evolution of the monsoon and determine its topological characteristics (Satyam 1988).

The data was taken from the 116-year record of the all-India summer monsoon (June to September) rainfall for the period 1871 to 1986. The correlation function for the time series was computed beginning in a two-dimensional phase based on Grassberger and Procaccia method (1983). The strange attractor dimension was found as 5.1.

9.5.3 Dynamical weather systems

Tropical cyclones, extra tropical cyclones (which affect the Indian region as Western Disturbances with diffuse frontal characteristics) and monsoon depressions (which typically occur in the Bay of Bengal and Arabian Sea during the Southwest Monsoon) are the examples of the dynamical weather systems which are associated with rain (sometimes extensive and heavy) and strong winds. Of these, tropical cyclones cause large casualties besides considerable damage to property and agricultural crop near the coastal districts. The maximum destruction is generally within 100 km from the centre of the cyclone caused by fierce winds, torrential rain, flooding and high storm tide due to the combined effect of storm surge and tide. These storms form out of a weak low pressure system over warm seas where moisture provides the chief source of energy for the development of cyclonic storm.

The strange attractor dimension for typhoons in Pacific ocean was found as 4.185 indicating 5 variables in the dynamic system (Yongqing and Shaojin 1994). Pal (1991) found the attractor dimension for the tropical cyclones over the north Indian region. His results based on 105 cyclones during October and November gave the strange attractor dimension between 5 and 6.

9.5.4 El Nino – Southern Oscillation (ENSO)

The phenomenon of El Nino has attracted the attention of the scientists due to its influence over the global weather. They occur irregularly every 2 to 6 years and are manifested through extended periods of anomalously warm sea surface temperature off the coast of South America. These changes are intimately linked to the atmospheric zonal circulation in this region known as Southern Oscillation. The zonal transport of atmospheric mass (Walker circulation) across the Pacific ocean produces a dipole surface pressure fluctuation which is clearly identified by subtracting monthly average sea-level pressures at Darwin from similarly averaged values at Tahiti. Persistent negative values of this Southern Oscillation Index (SOI) correlate well with El Nino events. Seasonal anomalies also occur in sea surface temperatures. Normalized monthly values of SOI beginning in January of 1882 and continuing through April of 1992 (1324 consecutive months) were used to study the predictability of this phenomenon.

Bauer and Brown (1992) using the method of singular-spectrum analysis (SSA) showed that the underlying dynamics of the ENSO system can be captured in a deterministic low order model.

Elsner and Tsonis (1992) adopted another procedure called the method of surrogate, which generates a large number of random sequences of equal length as the time series to be tested. The idea is that the surrogate time series should be a non-deterministic record but similar in appearance to the original data. In this process, the autocorrelation is preserved in the surrogate data, which is otherwise random. A null hypothesis is defined against which the raw data can be tested using a discriminating statistics. According to the above algorithm for generating surrogate records, the null hypothesis is that the raw data come from a linear autocorrelated Gaussian process. The discriminating statistics (e.g., Lyapunov spectrum, correlation dimension etc.) is computed for each surrogate time series and this distribution approximated. If the discriminating statistics for the real data is significantly outside the range of the distribution based on the surrogates, then the null hypothesis of linearly correlated noise is rejected. It can, therefore, be concluded that significant nonlinear structure is present in the record. This technique further provided evidence of low dimensional chaos in ENSO (Elsner and Tsonis 1993).

The El Nino's dynamics is best explained through the equatorial ocean using Kelvin and Rossley waves. Accordingly, their prediction is based on Atmosphere – Ocean coupled models. Tziperman et al. (1997) suggested an alternate method for controlling spatio-temporal chaos in realistic El Nino prediction model. A criterion was evolved for determining the optimum points in reconstructed delay–coordinate phase space to apply the feedback control. This led to stabilization of a full domain oscillation in an unstable periodic orbit through a single degree of freedom at a carefully single 'choke point' in space. Physically, the 'choke point' occurs at the western boundary of the Pacific ocean, which affects the entire tropical region through the reflection of the Rossby waves into Kelvin waves. Thus the control variable was taken as the Kelvin mode amplitude at the western boundary with a single degree of freedom out of thousands in the model.

9.5.5 Radio refractivity

Radio refractive index (RRI) at the ground surface has been extensively used to predict microwave signal strengths in different parts of the world. It is based on three meteorological variables namely atmospheric pressure, temperature and water vapour.

Surface meteorological observations and the radio sonde profiles provide the basic data for modeling. Direct measurements of RRI through microwave refractometers are costly and can be limited to only a few places.

Of late, the applications of the underlying principles in the propagation of microwaves in the earth atmosphere system are numerous. Global Positioning Systems (GPS) are now extensively used to measure crustal deformation, plate movements, and retrieval of water vapour in the atmosphere and soil moisture for agricultural operations.

However, the accuracy of microwave measurements is based on the profiles of radio refractivity and the calibration procedures. To validate the results, the effect of regional models (Srivastava and Pathak 1970) on the accuracy of GPS measurements needs to be examined for Indian region.

The radio refractivity at microwave frequencies is complex and can be expressed in two parts as,

$$N = N_d \,(\text{Dry part}) + N_w \,(\text{Wet part}) \tag{9.9}$$

Radio refractivity (N) was calculated based on daily values using the formula given below

$$N = \frac{77.6p}{T} \left[1 + \frac{7.71m}{T} \right] \tag{9.10}$$

where T is dry bulb temperature (^{0}K), m is mixing ratio (gm/kg), p is station level pressure (hPa).

Daily mixing ratio m for each station was computed from

$$m = \frac{0.622e_s}{P - e_s} \tag{9.11}$$

where e_s is saturation vapor pressure (hPa) given by

$$e_s = \sum_{n=0}^{6} a_n T_d \qquad (9.12)$$

The dew point temperature T_d (°C) and constants a_0 to a_6 are given below

$a_0 = 6.107799961 \times 10^{-0}$, $a_1 = 4.426518521 \times 10^{-1}$ (9.13)

$a_2 = 1.428945805 \times 10^{-2}$, $a_3 = 2.650648471 \times 10^{-4}$

$a_4 = 3.031240396 \times 10^{-6}$, $a_5 = 2.034080948 \times 10^{-8}$

$a_6 = 6.136820929 \times 10^{-11}$.

Large spatial and temporal variations of the radio refractivity in the atmosphere are attributed to the water vapor, which introduces 'delay' due to the absorption or attenuation of microwaves caused, by the complex part of the radio refractivity.

The correlation dimension of attractors of radio refractivity is obtained from the correlation integral C_m (r) for different values of r using Grassberger and Procaccia (1983) method. $\ln C_m$ (r) is plotted against $\ln$(r) in Figs. 9.3 (a-c) for $\tau = 7$ day. To obtain strange attractor dimension (D) using Eq(9.4) we require the slope d of the straight line passing through the points corresponding to each embedding dimension m. However, C_m(r) saturates at large values of r due to finite size of the attractor and at small values of r due to finite N.

The computations were made for 2 to 18 embedded dimensions. The results are shown in Figs. 9.4 (a-c) for three representative stations namely, Gauhati, New Delhi and Mumbai corresponding to three distinctive radio climatic regimes. As given in Table 9.2 and Fig. 9.4 (a-c) there is saturation for $\tau = 3$, 5 and 7 days but only representative curves for $\tau = 7$ are given in the Figs. 9.4 (a-c).

The strange attractor dimension for Gauhati, Nagpur, New Delhi, Mumbai, Calcutta, Madras, Port Blair and Vishakhapatnam are given in Table 9.2. It may be noted that the strange attractor dimension in the Indian region lie in the range of 5.9 to 8.3 implying that 6 to 9 parameters are needed for modeling the radio refractivity. Among the inland stations, only 7 parameters are needed for modeling at Nagpur and Gauhati while at New Delhi, eight parameters are needed. On the other hand, at a coastal station like Mumbai where anomalous propagation extending up to coast of Arabian Sea was reported since World War II, only six parameters are needed for modeling. This could be attributed to the different physical processes for the ducting conditions over the inland stations as compared to that over sea.

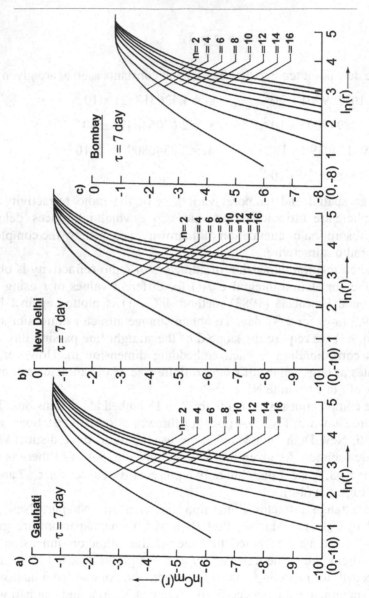

Fig. 9.3 lnCm(r) versus ln(r) for: (a) Gauhati, (b) New Delhi and, (c) Bombay (after Srivastava et al. 1994)

At Vishakhapatnam, which has the highest value of surface refractivity among coastal stations during the southwest monsoon, nine parameters are needed to model the refractivity system. Such a large difference in the modeling parameters between Mumbai and Vishakhapatnam may be at-

tributed to greatest turbulence in the refractivity parameter, which makes
the system more complex.

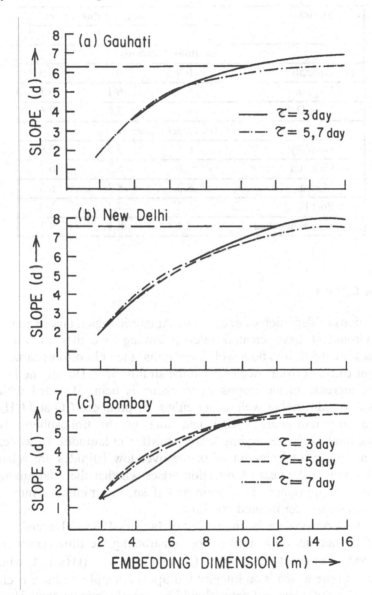

Fig. 9.4 Slope (d) versus embedding dimension (m) for: (a) Gauhati, (b) New
Delhi and (c) Bombay

Table 9.2 Strange attractor dimension over India (Radio Refractivity)

Station	Strange attractor dimension		
	$\tau = 3$	$\tau = 5$	$\tau = 7$
(a) Inland stations			
Gauhati	6.9	6.3	6.3
Nagpur	6.6	6.1	6.1
New Delhi	8.0	7.6	7.6
(b) Coastal stations			
Bombay	6.5	5.9	5.9
Calcutta	7.0	6.5	6.3
Madras	8.3	8.3	8.3
Port Blair	7.3	7.3	7.3
Vishakhapatnam	8.1	8.1	8.1

9.5.6 Ozone

Observations of depletion of ozone over Antarctic region and its impact on the environment have created interest among scientists to know the processes involved. It is now well known that the chlorofluorocarbons are the major cause of ozone depletion in the stratosphere. During the last few decades, increase in the tropospheric ozone is being detected which is attributed to industrial as well as agriculturally produced CO and CH_4. The decrease in stratosphere ozone and increase in tropospheric ozone influence total ozone in varying degree at different latitudes. However, the variations in the total amount of ozone over low latitudes are relatively small throughout the year. A question arises whether the total amount of ozone in the atmosphere is chaotic and if so, the minimum number of parameters needed for its predictability.

Daily total ozone data for the period of January 1975 to December 1993 at 5 Indian stations were utilized for constructing the time series x(t). A phase-space was constructed with x(t),x(t+τ),..........,x(t+(n-l) τ),where τ, is the delay time which is an integral multiple of sampling time τ is chosen such that the shifted co-ordinates should be linearly independent. The largest Lyapunov exponent was found to be positive for all five stations, which indicate that variation of total ozone in the atmosphere is chaotic i.e. the evolution of the system lies on a strange attractor. The plot of $\ln C_m(r)$ vs. $\ln(r)$ is shown in Fig. 9.5a. The slope d is calculated from the straight line

passing through the points corresponding to each embedding dimension
.The plot of slope d versus embedding dimension m is shown in Fig. 9.5b.
Table 9.3 gives the strange attractor dimension of total ozone for all the
five stations in India. The non-integral dimensions obtained suggest that
the system is chaotic and the trajectories of the total ozone lie on a strange
attractor. The values of strange attractor dimension at the stations namely
Varanasi, Srinagar, Kodaikanal and Pune lie between 5.5 and 5.7, which
indicate that atleast six parameters are needed to model the total atmosphere ozone. It is also seen that saturation of the slope takes place at embedding dimension 14, which is an upper bound for the number of variables sufficient to model the dynamics of the attractor. The strange
attractor dimension for New Delhi was found to be 6.5 indicating that atleast seven parameters are needed to model total atmospheric ozone over
New Delhi. However, the saturation of the slope took place at embedding
dimension of 14 in this case also, suggesting that the upper bound of the
number of parameters sufficient to model the dynamics of the attractor is
the same for all the stations.

Table 9.3 Strange attractor dimension of total ozone over Indian stations

S.No	Station Name	Period		No. of data points	Strange attractor dimension
1.	Varanasi	January 1976 December 1992	to	6210	5.7
2.	Srinagar	January 1976 December1989	to	5114	5.7
3.	New Delhi	January 1983 December1988	to	2192	6.5
4.	Kodaikanal	January 1975 December1991	to	6209	5.7
5.	Pune	January 1980 December1993	to	5114	5.5

9.5.7 Ionospheric scintillations

Spatial variations of electron density in the ionosphere cause scintillations
in the amplitude and phase of radio wave propagating through the ionosphere. Stochastic approach adopted earlier in the theories of ionospheric
scintillation due to complexity of the electron density structures was found
to be chaotic with low dimension of 4.41(amplitude data) and 3.61 (phase

data) for the strange attractor. This study was based on the amplitude and phase measurements carried out on a 140 MHz signal transmitted from the geostationary satellite ATS-6 and received in India at Ootacamund using 800 data points (Bhattacharya 1990).

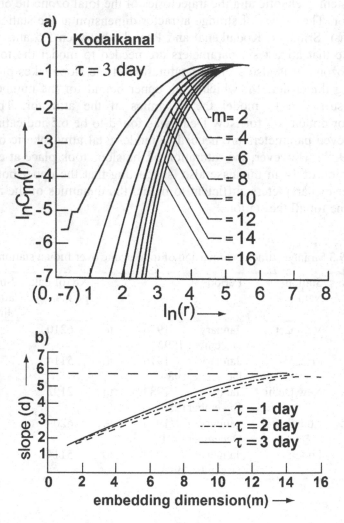

Fig. 9.5 (a) Daily total ozone amount (January 1975 – December 1991). Distance dependence of the correlation function for a sequence of embedding dimensions (2,4,8,12,14 and 16) (b) Dimensionality d of the attractor function of embedding dimensions (January 1975 December 1991). (after Srivastava and Sinha Roy 1997)

Magnetosphere –Ionosphere dynamics

The Earth's atmosphere extends up to the magnetopause, which acts as a boundary between the Earth and the interplanetary medium. The geomagnetic field deflects the solar wind (which is an ionized and highly conducting gas, consisting mainly of electrons and protons, being emitted continuously from the sun) resulting in the formation of a cavity, which is known as the magnetosphere. The boundary of the magnetosphere is called the magnetopause. As a result of solar wind geomagnetic field interaction, the geomagnetic field is compressed on the day side (to ~$10R_E$) and stretched on the night side into a long tail extending by 1000 R_E, where R_E is the Earth's radius. Further, magnetosphere extracts energy from the solar wind continuously and dissipates it by setting up a complex pattern of several current systems .The solar wind power which penetrates the magnetosphere is 10^{10} to 10^{11} W during geomagnetic quiet time and 10^{12} W during geomagnetically disturbed periods.

The geomagnetic indices give a measure of the large scale eastward (AU) and westward (AL) electrojet intensities. It is believed that AL, AU and auroral electrojet index (AE) quantify the response of the magnetosphere to the solar wind during the sub-storms. The field-aligned currents that connect the magnetotail to the auroral zone are closed by the high latitude ionospheric electrojets. The index AL/AU characterizes the fluctuations in the westward/eastward electrojet field, while the index AE=(AU–AL) is a measure of both electrojets. Chaotic techniques were applied to analyse the AE or AL time series to understand the sub-storm dynamics. The geomagnetic index (AE or AL time series) was used to construct higher dimensional phase spaces in an attempt to find the low-dimensional response of the magnetosphere. Several scientists have adopted different methodology to study the problem. Vassiliadis et al. (1992) used AE data to study chaoticity of the geomagnetic activity based on observations for 21 days in January 1983 averaged over 1-min interval in 5000 point subsets. Each data subset was embedded in a reconstructed state space using the method of time delays. It was found that the dimension of the attractor determined from was about 3.6 on average, with the dimension being independent of activity level. Baker et al (1991) found the correlation dimension as 4.0 using 40000 data sets. However, lower strange attractor dimensions of the order of 2.5 were also reported. Sharma et al (1993) analyzed the AE data by using singular spectrum analysis and found that the sub-storm attractor persists but with dimension ~2.5. Takalo et al (1993) showed that many properties in AE data are similar to those of bicolored noise. They found average correlation dimension of 3.4 by analyzing 1 minute AE data from the years 1979-85. The phase space reconstruction using the singular spectrum analysis brought out a fractal dimension of 2.5. Vassiliadis et al (1992) subjected 2.5 min resolution AL time series of length N=29,000 to the time delay embedding techniques with τ = 175 min and m=10 to determine the Lyapunov exponent. They found the Lyapunov exponent to be 0.11 ± 0.05 min^{-1}, which supported the chaotic aspects of the sub-storm activity.

The loading–unloading model of geomagnetic activity generally consists of the loading phase in which the magnetotail field increases due to the day side reconnection, and the unloading phase in which the stored magnetic energy is suddenly released. This model may reflect interval magnetosphric dynamics through a low dimension of attractor. This appears more plausible because the directly driven model does not involve triggering or sudden release of energy and would imply an attractor of high dimension.

9.6 Prediction aspects

It would be seen from the above results that several components in Earth system show chaotic behaviour. Among these, the strange attractor dimension in atmospheric pressure, minimum and maximum temperature, monsoon rainfall, cyclones and ozone remained of the same order namely 5 to 6. It implies that atleast 6 to 7 parameters are required for predictability in general. Somewhat larger variations are found in radio refractivity in the atmosphere. In the case of magnetosphere–ionosphere dynamics, number of parameters is less, which is of the order of 3 to 4. In the case of earthquakes, there is wide regional variation. For the Himalayan region, this number of parameters for predictability is larger than for the peninsular India. In Japanese region, the strange attractor dimension for earthquakes is of the same order as volcanic eruptions in Hawaii.

The predictability h on attractors can be estimated from slopes of the distribution function in a $\ln C_m(r)$ vs. $\ln(r)$, if the dimension m is chosen sufficiently high that the attractor is embedded in the phase space of time shifted coordinate (Fraedrich 1987). The difference between $\ln[C_{m+k}(r)]$ and $\ln[C_m(r)]$ at a fixed distance threshold leads to the mean predictability as

$$h \approx \frac{1}{\tau K \ln \dfrac{C_m}{C_{m+h}}} \tag{9.14}$$

The fixed distance should be selected from the ln(r) interval where the related distribution ln[C$_{m(}$ r)] can be approximated by straight lines of identical slopes, i.e. where C(r) ~r $^\infty$ is satisfied. The inverse value 1/h of the mean divergence defines a mean time scale up to which predictability may be possible, if e-folding volume expansion is considered. Using the above relationship, Srivastava and Sinha Ray (1997) found the predictability of station level pressure and temperature (maximum and minimum) in the Indian region as 8 to 11days. The time limit of predictability of typhoons in Pacific was reported as about 61 hours. Some attempts have been made to apply chaotic results for the prediction of tracks of tropical cyclones in Bay of Bengal and climate epochs.

9.6.1 Tropical cyclone track prediction

Although weather satellites have helped on the prediction of the track of cyclones, the numerical prediction techniques are being improved in this direction to forecast the future movements well in advance. Pal (1991) used deterministic chaos to supplement the efforts for the prediction of tropical- cyclone tracks.

The following prediction formula was suggested assuming that the future position of the cyclone depends on the previous position, at n+1 time step

$$X_{n+1} = \sum_{i=1}^{M} a_i X_{n+i-M} \quad Y_{n+1} = \sum_{i=1}^{M} b_i Y_{n+i-M} \qquad (9.15)$$

where X$_n$ and Y$_n$ are cyclone position coordinates and regression coefficients a$_i$ and b$_i$ were determined from past cyclones data for the period 1890 – 1970 published by the India Meteorological Department. This process is repeated for further prediction. The coefficient a$_i$ and b$_i$ are determined by least squares method using all seven consecutive places of 105 cyclone trajectories. More weight was given to the current position and nearby point than to the far away points. Out of the results presented for two tropical cyclones during November 1984 and 1988, the turning direction for the 1984 cyclone was successfully predicted which appeared to be failed if the persistence method was used (Fig.9.6).

9.6.2 Climate dynamics

(a) The important feature of climatic records is that of variability. Nicolis (1982) modeled climate according to a forced non-linear oscillation and showed the existence of complex non-periodic behaviour similar to that observed in climatic records. It is possible however that climate as a dynamical system may have multiple attractors. Its dynamics may be dictated by jumps from one attractor to another. Fraedrich (1987) examined the oxygen isotope (180) record of planetonic species from a 10.7m long deep sea core from the eastern equatorial Atlantic and found a predictability time scale of 10 to 15 thousand years with strange attractor as 4.4 to 4.8.

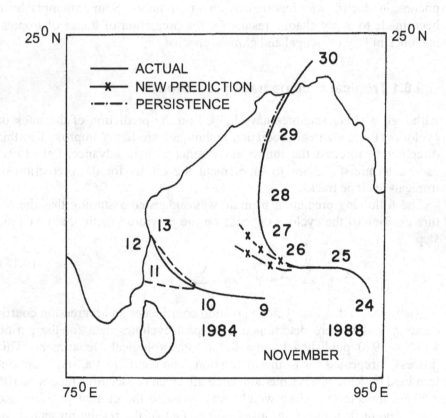

Fig. 9.6 Sample predictions for November cyclones of 1984 and 1988. Thick line is actual track and dashed lines with dots and crosses are predictions by persistence (Pal PK 1991)

(b) India Meteorological department has been issuing long-range weather forecast for the rainfall during June to September every year based on the 16 parameters since 1988, using power regression models. Although the predictions lie generally within the statistical limits of normal but in

2001, the model failed. New models have been evolved from April 2003 for long range weather forecasting in India. But the problems remain due to the large number of parameters in the model, which change correlations with time. Srivastava and Singh (1993) used the principal components analysis to explain the overall variability of rainfall using less number of parameters. Models based on these results need to be synthesized with dynamical approach. However, the concepts of compound chaos have so far been used for only two systems (Singh et al 1996). Srivastava (1997) reported that this may be a severe constraint in its application. Similar problems are envisaged if we wish to couple more than two dynamical systems components of the Earth system.

9.7 Discussion

It is seen from the above results that several components in Earth system show chaotic behavior. Among these, the strange attractor dimension in atmospheric pressure, minimum and maximum temperature, monsoon rainfall, cyclones and ozone remained of the same order namely 5 to 6. It implies that atleast 6 to 7 parameters are required for predictability in general. Somewhat larger variations are found in radio refractivity in the atmosphere. In the case of magnetosphere–ionosphere dynamics, number of parameters is less, which is of the order of 3 to 4. In the case of earthquakes, there is wide regional variation. For the Himalayan region, this number of parameters for predictability is larger than for the peninsular India. In Japanese region, the strange attractor dimension for earthquakes is of the same order as volcanic eruptions in Hawaii.

The predictability on attractors can be estimated from slopes of the distribution function in a $\ln C_m(r)$ vs. $\ln(r)$, if the dimension m is chosen sufficiently high that the attractor is embedded in the phase space of time shifted coordinate. The inverse value $1/h$ of the mean divergence defines a mean time scale up to which predictability may be possible, if e-folding volume expansion is considered. The time limit of predictability of typhoons in Pacific was reported as about 61 hours. Some attempts have been made to apply chaotic results for the prediction of tracks of tropical cyclones in Bay of Bengal and climate epochs.

It is of interest to note that several earthquakes are generally chaotic in nature. Thus, prediction efforts need to be intensified in the same way as for weather, which is also chaotic. In view of low strange attractors in the Koyna region, India with a localized source and recurrence of earthquakes of magnitude 5 from time to time, the methodology suggested by

Srivastava and Bhattacharya (1998) based on the Principal Component
Analysis offers a systematic approach to evolve a prediction strategy.

Acknowledgements

I am thankful to IMD, India for granting permission to reproduce figures
9.1, 9.3 and Table 9.1 from Mausam. I acknowledge Proceedings of National Academy of Sciences, India for their permission to reproduce figures 9.2, 9.5 and Tables 9.2, 9.3.

9.8 References

Baker DN, Klimas AJ, Roberts DA (1991) Examination of time-variable input effects in a nonlinear analogue magnetosphere model. Geophy Res Lett 18:1631–1634

Bauer ST, Brown MB (1992) Empirical Low– order Enso dynamics. Geophys Res Lett 19: 2055-2058

Beltrami H, Mareschal JC (1993) Strange seismic attractor? Pure Appl Geophys 141: 71-81

Bhattacharya SN, Srivastava HN (1992) Earthquake predictability in Hindukush region using chaos and seismicity pattern. Bull Ind Soc Earthq Tech 29: 23-25

Elsner JB, Tsonis AA (1993) Non–linear dynamics established in the EN-SO. Geophys Res Lett 20: 213-216

Elsner JB, Tsonis AA (1992) Non-linear prediction as a way of distinguishing chaos from random fractal sequences. Nature 358: 217-220

Fraedrich K (1987) Estimating weather and climate predictability on attractors. Jr Atmosph Sci 44: 722-728

Grassberger P, Procaccia I (1983) Characteristics of strange attractors. Phys Rev Lett 50: 346-349

Gutenberg B (1959) Physics of the Earth's interior. Academic Press, New York

Horowitz FG (1989) A Strange attractor underlying Parkfield seismicity EOS 70: 1359

Julian BR (1990) Are earthquakes chaotic? Nature 345: 481-482

Lowrie W (1997) Fundamentals of geophysics. Cambridge University Press, UK

Nicolis C (1982) Stochastic aspects of climatic transitions-response to a periodic forcing. Tellus 34:1-9

Pal PK (1991) Cyclone track prediction over the north Indian ocean. Mon Wea Rev Notes and Correspondence 119: 3095-3098

Polvos GP, Karakatsanis L, Latoussakis LB, Dialetis D, Papaisannou G (1994) Chaotic analysis of a time series composed of seismic events recorded in Japan. Intern Jr Bifurc & Chaos 4: 87-98

Rikitake T (1958) Oscillations of a system of disk dynamos. Proc Cambridge Phil Soc 54:89:105

Ruelle D (1990) Deterministic chaos: the science and the fiction. Proc Roy Soc Lond A-427: 241-248

Satyan V (1988) Is there a strong attractor in monsoon rainfall? Proc Indian Acad Sci EPS 97: 49-52

Sharma AS, Vassiliadis D, Papadopoulos K (1993) Reconstruction of Low- Dimensional Magnetospheric Dynamics by Singular Spectrum Analysis. Geophys Res Lett 20: 335

Singh R, Moharir PS, Maru VM (1996) Compound chaos. Int Jr Bifurc Chaos 6:383-393

Sornette A, Dubois J, Cheminee JL, Sornette D (1991) Are Sequences of Volcanic Eruptions Deterministically Chaotic? J Geophys Res 96:11931

Srivastava HN, Pathak RB (1970) Regional models of Radio atmosphere over India. Jr Inst Telecom Engrs 16:171-180

Srivastava HN, Singh SS (1993) Empirical orthogonal functions associated with parameters used in long range forecasting of Indian summer monsoon. Mausam 44: 29-34

Srivastava HN, Bhattacharya SN, Sinha Ray KC (1994a) Strange attractor dimension as a new measure of seismotectonics around Koyna reservoir India. Earth Plan Sci Lett 124: 57-62

Srivastava HN, Sinha Ray KC, Bhattacharya SN (1994b) Strange attractor dimension of surface radio refractivity over Indian stations. Mausam 45: 171-176

Srivastava HN, Bhattacharya SN, Sinha Ray KC, Mahmoud SM, Yunga S (1995) Reservoir associated characteristics using deterministic chaos in Aswan, Nurek and Koyna reservoirs. PAGEPOH 145: 209-217

Srivastava HN, Bhattacharya SN, Sinha Ray KC (1996) Strange attractor characteristics of earthquakes in Shillong plateau and adjoining regions. Geophy Res Lett 23: 3519-3522

Srivastava HN (1997) Compounding meteorological parameters in long range weather forecasting using deterministic chaos. IAMAP-IAPSO Symposium Melbourne Australia

Srivastava HN, Sinha Ray (1997) Predictability of geophysical phenomena using deterministic chaos. Proc Nat Acad Sci LXVII Part VI: 305-345

Srivastava HN, Bhattacharya SN (1998) Application of principal component analysis to some earthquake related data in the Koyna region, India. Eng Geology 50: 141-151

Srivastava HN, Sinha Ray KC (1999) Deterministic chaos in earthquake occurrence in Parkfield California region and characteristic earthquakes. Mausam 49: 104

Stewart CA, Turcotte DL (1989) The route to chaos in thermal convection at infinite Prandtl number 1 some trajectories and bifurcations. Jour Geophys Res 94: 13707

Takalo J, Timonen J, Koskinen H (1993) Correlation dimension and affinity of AE data and bicolored noise. Geophys Res Lett 20: 525-550

Tiwari RK, Rao KNN (2001) Power law random behaviour and seasonality bias of northeastern India earthquakes. J Geol Soc India 57: 369-376

Tiwari RK, Srilakshmi S, Rao KNN (2003) Nature of earthquake dynamics in the central Himalayan region: a non-linear forecasting analysis. J Geodyn 35: 273-287

Tziperman E, Saher H, Zebiak SE, Cane MA (1997) Controlling spatio temporal chaos in a realistic El Nino prediction model. Phy Rev Lett 79: 1034-1038

Vassiliadis D, Sharma AS, Papadopoulos K (1992) Low dimensionality of magnetospheric activity. EOS Trans AGU Spring Meeting Suppl 73: 270

Wolf A, Swift JB, Swinney LH, Vastano JA (1985) Determining Lyapunov exponents from a time series. Physica 16 D: 285-317

Yongqing P, Shaojin Y (1994) Seasonal variation features of Western North Pacific tropical cyclone tracks with their predictability. Advances in atmospheric sciences 11: 463-469

Index